# Hands-on AIOps

## Best Practices Guide to Implementing AIOps

Navin Sabharwal
Gaurav Bhardwaj

Apress®

# Hands-on AIOps: Best Practices Guide to Implementing AIOps

Navin Sabharwal
New Delhi, India

Gaurav Bhardwaj
New Delhi, India

ISBN-13 (pbk): 978-1-4842-8266-3
https://doi.org/10.1007/978-1-4842-8267-0

ISBN-13 (electronic): 978-1-4842-8267-0

Managing Director, Apress Media LLC: Welmoed Spahr
Acquisitions Editor: Celestin Suresh John
Development Editor: Laura Berendson
Coordinating Editor: Shrikant Vishwakarma
Copyeditor: Kim Wimpsett

Cover designed by eStudioCalamar

Cover image by Pixabay (pixabay.com)

Distributed to the book trade worldwide by Apress Media, LLC, 1 New York Plaza, New York, NY 10004, U.S.A. Phone 1-800-SPRINGER, fax (201) 348-4505, e-mail orders-ny@springer-sbm.com, or visit www.springeronline.com. Apress Media, LLC is a California LLC and the sole member (owner) is Springer Science + Business Media Finance Inc (SSBM Finance Inc). SSBM Finance Inc is a **Delaware** corporation.

For information on translations, please e-mail booktranslations@springernature.com; for reprint, paperback, or audio rights, please e-mail bookpermissions@springernature.com.

Apress titles may be purchased in bulk for academic, corporate, or promotional use. eBook versions and licenses are also available for most titles. For more information, reference our Print and eBook Bulk Sales web page at www.apress.com/bulk-sales.

Any source code or other supplementary material referenced by the author in this book is available to readers on GitHub (https://github.com/Apress). For more detailed information, please visit www.apress.com/source-code.

Printed on acid-free paper

*Dedicated to family, friends, mentors, and Almighty God who has been there right from the beginning, giving strength and skill.*

# Table of Contents

# About the Authors

**Navin Sabharwal** has more than 23 years of industry experience. He is a thought leader and head of strategy and development in the areas of AgileOps, cloud computing, DevSecOps, AIOps, FinOps, artificial intelligence, and SaaS product engineering. He holds seven patents in the areas of AI and machine learning. He has authored 15+ books in these domains in the areas of cloud capacity management, observability, DevOps, cloud automation, containerization, Google AutoML, BERT, and NLP. He can be reached at Navinsabharwal@gmail.com and www.linkedin.com/in/navinsabharwal.

**Gaurav Bhardwaj** is a seasoned IT professional and technology evangelist with more than 17 years of expertise in service assurance, enterprise monitoring, event management, cloud computing, AI/ML-based software product development, engineering, and data analytics. He has international experience in solution designing and execution of IT automation strategies that are aligned with business goals. Gaurav has a proven track record of achievements in the roles of enterprise architect and consultant on IT transformation projects for large and complex global engagements (including multiple Fortune 500 companies), migrating from legacy processes and platforms to next-generation IT environment powered by cloud-native and containerized apps, SDI and AIOps, and DevOps methodologies. He can be reached at Gaurav.Bhardwaj22@hotmail.com and www.linkedin.com/in/gauravbhardwaj-1.

# About the Technical Reviewer

 **Pradeepta Mishra** is the head of AI at FOSFOR by L&T Infotech (LTI), leading a group of data scientists, computational linguistics experts, and machine learning and deep learning experts in building artificial intelligence products. He was awarded "India's Top 40Under40DataScientists" by *Analytics India Magazine* for two years in a row, 2019 and 2020. As an inventor, he has filed five patents in different global locations. He is the author of five books, and his first book has been recommended by the HSLS Center at the University of Pittsburgh, Pennsylvania. His fourth book, *Pytorch Recipes*, was published by Apress and added to the Buswell Library in Illinois. His fifth book, *Practical Explainable AI Using Python*, was recently published by Apress and has been recognized as a textbook by the Barcelona Technology School (BTS) for its big data analytics course. He delivered a keynote session at the Global Data Science conference in California in 2018 and has delivered 500+ tech talks on data science, ML, DL, NLP, and AI in various universities, meetups, technical institutions, and community-arranged forums. He is a visiting faculty and academic advisory board for AI/ML at Reva University in Bangalore, India, as well as at various other universities. He has mentored and trained more than 2,000 data scientists and AI engineers in the past decade.

# Acknowledgments

To my family, Shweta and Soumil, for being always there by my side and letting me sacrifice their time for my intellectual and spiritual pursuit. This and other accomplishments of my life wouldn't have been possible without your love and support.

To my mom and my sister for your love and support as always: without your blessings, nothing is possible.

To my coauthor Gaurav and the team at HCL who have been a source of inspiration.

Thank you to Celestine and the entire team at Apress for turning our ideas into reality. It has been an amazing experience authoring with you, and over the years, the speed of decision-making and the editorial support has been excellent.

—Navin

To my mom and dad for always being there by my side, giving invaluable support and encouragement and continuously striving for knowledge and wisdom.

To my wife and lovely daughters, Vaidehi and Maanvi, for being inspirations to me and for sacrificing their time.

To my coauthor and mentor, Navin, thank you for your guidance, support, and feedback.

To my teammates, Vipul Tyagi and Rajesh Upadhyay, thank you for your technical support.

Thank you, Lord, for giving me wisdom and strength through every aspect of life.

—Gaurav

# Preface

Before starting our AIOps journey, let's briefly discuss automation and how it has evolved over the last decade.

Automation in the technology domain is defined as a system where a process or task can be performed with minimal human supervision and action. Humans have been automating tasks forever, and eventually mechanical machines that reduced human effort and increased efficiency were invented to reduce human effort. Today things such as manufacturing, which were partly automated earlier, are moving to 3D printing, which completely automates the process of manufacturing; however, designing what to manufacture is still in the human domain.

Thus, automation reduces human effort and uses machines or software to complete definable, repeatable tasks.

In the IT domain, there are various tasks that humans perform, including envisioning a new product or application, developing the software that translates these requirements into working software, and deploying infrastructure and applications and keeping them updated through their lifecycle.

IT teams have used automation extensively in every area, from software development to operations; however, this has largely been siloed and done without a formal system or method to automation. IT teams have used scripts, runbook automation tools, job scheduling systems, and robotic process automation systems to automate their tasks. These tools have resulted in increased efficiency and reduced the human requirement to operate IT environments.

With the increased adoption of cloud computing and DevOps principles, the provisioning of infrastructure-as-a-service and platform-as-a-service environments has also been automated, as has the deployment of applications. This has resulted not just in automating the tasks and increasing efficiency but also in agility and speed, which provides businesses with the support to pivot and adapt to changing market needs by quickly changing the functionality and features based on customer and market feedback.

IT runs on three pillars: process, people, and technology. To be able to automate, one needs to be aware of the inter-relationships between these pillars. People use defined processes to work on technology, and with automation we are essentially automating the current processes that people are using to operate an environment. However, with the changing technology landscape and increased adoption and maturity of AI and machine learning capabilities, we can now look at the current processes and formulate new ones to leverage the transformational capabilities provided by these technologies. The current processes were set up with state-of-the-art technology at that point in time, and these processes then defined how humans should operate within that process to execute tasks to accomplish a goal; however, with a drastic change in the technology landscape, the processes need to change and adopt. As an example, with the cloud becoming prevalent, the IT processes need to change and adapt, and the sequential, nonautomated processes need to change to cater to the new capabilities such as autoprovisioning. In 2013, I talked about how capacity management would drastically change in the cloud computing world and that new procedures for cloud cost management would be required. Some of these concepts have been expanded to cover the entire financial operations piece under the umbrella of *cloud FinOps*.

Similarly, IT operations automation had existed in siloes for all this while. People used scripts, monitoring tools, runbook automation, configuration and deployment automation, and RPA tools and automated service management processes using ITSM tools. However, all that was happening under different domains, all getting integrated with siloed integrations.

On the technology front, AI and machine learning technologies became mainstream and were being used heavily in all aspects of business-facing and customer-facing applications from websites to search engines to collaboration and communications tools. AI took over the world of IT quickly; however, IT was late to adopt these technologies. While IT teams were using these technologies to create new applications with AI capabilities to customers, their own internal systems were still using older technologies and worked on processes and systems that had largely remained untouched for the past decade.

Through the DevOps and Agile movements, that transformational change had already transformed the way applications were being built, tested, and deployed, and most of the tasks in the development value stream were automated and integrated, resulting in organizations moving up to continuous delivery and continuous deployment. Similar to the transformation seen with DevOps, with AIOps gaining traction, we are going to witness a transformational change that will drastically change the way IT operations has been run in the past. Old processes, systems, tools, and ways of working will give way to the AIOps way of operating and help realize the vision of NoOps, where operations work seamlessly without disruption in an automated way without human supervision and intervention.

Enterprises today are at various levels of maturity when it comes to automation. Most are yet to achieve a high level of maturity in automation, partly because changing processes, breaking down walls of organization structures, and deploying new technology are complex, time-consuming tasks. In some organizations, where digitization and cloud computing programs are a part of large-scale transformation; AIOps and associated technology and process changes are enabling a complete transformation of digitization. One key factor in adopting AIOps has been a lack of comprehensive process and technology guidance in this domain. There is limited guidance available, and most of it is focused on products that vendors are trying to sell as a one-stop solution, which will leapfrog an organization to the next level. With this publication we are aiming to provide hands-on pragmatic guidance on how an organization can adopt these changes in the origination and what the pitfalls are to be avoided. Implementing AIOps is not about deploying an event correlation tool; it is about infusing AI and algorithms and automating all aspects of operations.

Most organizations today are at a level where they have automated individual tasks and have automated certain processes. Some have also achieved automation of complex end-to-end processes. However, an end-to-end self-sustaining model for automation is still missing. Automating a few processes versus automating everything that can be automated is a change that people haven't yet initiated. Automating a few processes and then scaling the automation across teams, functions, and departments is still work in progress in most organizations.

AIOps is not just a technology or process change; it is a cultural change where humans and AI work together in tandem supporting and augmenting each other's capabilities to achieve efficiencies and scalability, which IT teams have only wished for but have been unable to achieve. The promise of NoOps has failed to materialize because organizations have not been able to take a holistic and end-to-end approach to AIOps and instead have relied on quick fixes in the form of siloed tools that are deployed.

In enterprises that have moved up the maturity curve in automation, most activities that can be automated fully or partially are automated, and AI drives decision-making rather than humans using a dashboard and analytics to arrive at the next steps. It is a mindset change where humans have to accept the supremacy of the AI systems in certain areas and let go of the control mindset inherent in our species. IT teams are used to certainty and rule-based systems that have dominated the IT landscape for decades. AI-based systems are probabilistic and not definitive; thus, some enterprises have deployed AIOps systems but configured them in a manner where they are nothing but rule-based systems. Accepting probabilistic systems rather than step-by-step rule-based systems is a mindset change, and one needs to accept the risks that come with implementing such systems and controls to ensure best practices for running AI-based systems; human control on some of the actions is incorporated as part of the process changes.

Enterprises are also evolving to change their operating models to align with Agile principles; they are starting to integrate applications and infrastructure teams and are adopting new practices such as site reliability engineering to make operations support highly reliable and high-availability mission-critical systems and Internet-scale applications. AIOps is a key pillar in site reliability engineering–based operations. The book *Hands On Guide to AgileOps* published in 2021 by Apress provides hands-on guidance on how to adopt Agile for IT operations.

This book will take you through all aspects of AIOps including AI and machine learning algorithms and demonstrate how some of the features in AIOps are implemented in enterprise AIOps platforms. You will be able to leverage the capabilities provided by AIOps processes and platforms and take your organization to a higher level of maturity in operations. We will continue to provide updated content and guidance on the companion website, `www.AgileInfraOps.com`, where you will find articles and best practices as processes and technology continues to evolve.

# Introduction

This book is a hands-on guide to understanding AIOps in detail and assisting in implementing its technologies and processes in the organization.

The book explains the IT industry's need for AIOps, its architecture, and the technologies that contributed to its evolution. Readers will be able to grasp the core theoretical concepts around AIOps as well as go deeper by implementing hands-on examples and exercises that leverage machine learning techniques to implement AIOps. The book also provides guidance on how to set up AIOps in an enterprise and what pitfalls to avoid to complete a successful implementation. The book also explains the role of AIOps in the SRE and DevOps model and provides a detailed explanation of the enablement of key SRE principles by AIOps.

This hands-on book also provides an implementation of multiple AIOps use cases and provides sample code that you can run to better grasp some of the underlying principles and technologies that form the core of AIOps.

This book shares the best practices, processes, and guidelines to establish AIOps practice and systems in enterprises and the methods to measure the outcomes and continuously evolve.

# CHAPTER 1

# What Is AIOps?

This chapter introduces artificial intelligence for IT operations (abbreviated as AIOps). In today's rapidly transforming application and infrastructure landscape and adoption of cloud-native technologies, organizations are finding it difficult to provide 24/7 operations that can scale and meet the needs of businesses that now want much higher availability and agility to change based on customer and market feedback. This chapter also provides details on the benefits that AIOps brings to the table and how it supports the digitization journey of enterprises.

## Introduction to AIOps

AIOps is a buzzword in the operations world and was coined by Gartner in 2016. As mentioned, it means implementing artificial intelligence for IT operations. AIOps refers to a transformational approach to running operations using AI and machine learning technologies in various operations domains such as monitoring, observability, event correlation, service management, and automation. With the exponential growth seen in application and platform diversity, including the movement to microservices and cloud architectures, there is an enormous amount of data being generated in operations. The operations teams are overwhelmed with this vast amount of data and the diversity in applications, platforms, and infrastructures in the environment. Most enterprises today are rapidly migrating and adopting new technologies such as cloud and microservices architecture, and thus the rate of

© Navin Sabharwal and Gaurav Bhardwaj 2022
N. Sabharwal and G. Bhardwaj, *Hands-on AIOps*,
https://doi.org/10.1007/978-1-4842-8267-0_1

change in infrastructure and platforms is unlike anything seen before. The challenge in IT operations is to run steady-state operations without disruption and also support this agility and migration and bring new services into operations. These disruptions and changes are putting an enormous strain on the operations teams. Processes and systems that have worked in the past are not working anymore, and the new digitalized world with rapid changes both in applications and infrastructure is resulting in newer challenges. Thus, AIOps has evolved over the last few years as a potential solution to the operational challenges of the new model.

The huge amount of data getting generated from monitoring and observability systems is one of the sources of data that is fed into AIOps-based systems, and then AI and machine learning techniques are used to make sense of the data and filter the noise from critical events. This results in automating most of the tasks that were manual before and that relied on human judgment and tribal knowledge. The events that can cause disruption in business operations and are the root cause are efficiently identified using analytics techniques and thus provide immediate notification to the groups that are resolving them. Without AIOps this process will be difficult to run with changes in technology happening at a rapid pace; relying on older systems and tribal knowledge would mean operations is not scalable and predictable.

All of this is enabled by the emergence and maturity of artificial intelligence and machine learning technologies, which are the foundation of AIOps.

Artificial intelligence has transformed the way systems are developed and business processes are run. AI is everywhere from the image processing in your phone to recommendation engines on Amazon that provide you with new product recommendations based on your preferences. Face recognition and image beautification on phones are examples of applications that people are consuming every day without even knowing that artificial intelligence is powering these applications. Natural language processing advancements have transformed the way we interact with applications. Today voice assistants like Alexa, Siri, and Cortana are changing the way we communicate with content.

Information technology has leveraged these technologies to solve varying business problems in the areas of building recommendation systems, predictive systems, image recognition, voice recognition, text extraction, and natural language understanding systems.

However, when it comes to solving IT problems using artificial intelligence, enterprises and technology companies have yet to embrace this fully.

Finally, the AI technologies that IT teams have used to deliver exciting new applications for consumers and businesses are now finding their way into monitoring and managing the IT technologies. Thus, a new class of systems that are using algorithms to run IT operations was born.

AIOps is a term Gartner invented to describe a general trend of applying AI techniques to IT operations data sources to provide additional insights. AIOps is essentially a feature or set of features to analyze, combine, and collect data.

According to Garner, "By 2023, 40% of DevOps teams will augment application and infrastructure monitoring tools with artificial intelligence for IT operations (AIOps) platform capabilities." AIOps platforms are platforms that "utilize big data, modern machine learning and other advanced analytics technologies to directly and indirectly enhance IT operations functions with proactive, personal and dynamic insight."

Figure 1-1 defines the various areas in IT operations that are included in AIOps. These include monitoring, event analytics, predictive and recommendation systems, collaboration and engagement, and reporting and dashboarding technologies.

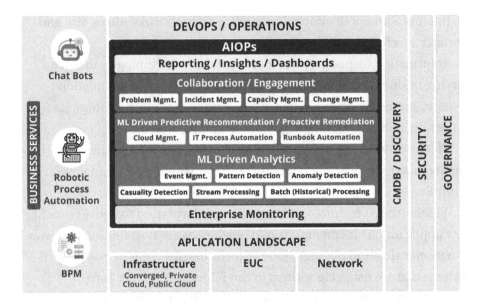

***Figure 1-1.*** *AIOps in the IT operations landscape*

AIOps covers various layers of information technology. From the network to the endpoints, everything in IT can use AIOps technologies to reap the benefits that AIOps provides.

Enterprise monitoring provides a real-time data feed to the AIOps system to perform ML-driven correlation and analysis using various techniques to detect patterns and anomalies, as well as perform causal-impact analysis. This is one of the most important stages as this analysis needs to consider both real-time streaming data as well as historical data to provide predictive recommendations or proactive remediations, which then get executed using IT process or runbook automation tools. The recommendations for resolution help the organizations in achieving end-to-end automation by resolving problems without human intervention.

The reporting and dashboard layer provides views for different IT teams and stakeholders to collaborate and manage incidents, capacity, change, and problems to further support business by providing KPIs and SLAs that are driven by insights and provide an element of predictive analytics to make operations more proactive.

AIOps systems leverage a Configuration Management Database (CMDB) to improve the quality of correlations and accuracy of predictions and recommendations, but organizations usually struggle in maintaining the accuracy of the CMDB and discovery data due to the ever-changing infrastructure landscape. With cloud computing, it is practically impossible to update the CMDB using traditional tools and processes. An AIOps system solves this problem by automatically populating the required missing data in the CMDB. This well-oiled engine of AIOps has to work within organizations' security policies defined under their governance framework. Various compliance needs like GDPR, data classifications, etc., should be considered at each layer of the AIOps engine while they are being integrated or set up. Once AIOps systems get up and running, they learn patterns, anomalies, and behavior using data over a period of time. Gradually, based on maturity, the AIOps system gets consumed by other technology or business units such as ChatOps, robotic process automation, business process automation, etc. More complex business process workflow or chat responses can be triggered based on the AIOps system recommendations.

Just like in software engineering, continuous integration and continuous deployment integrates different activities of development, testing, and deployment of applications and shares feedback for improvement. Similarly, AIOps is something that provides seamless integrations between various operations components and provides feedback for continuous service improvement.

The basic definition of AIOps is that it involves using artificial intelligence and machine learning to support all primary IT operations. As depicted in Figure 1-2, there are three layers in AIOps when it comes to event correlation.

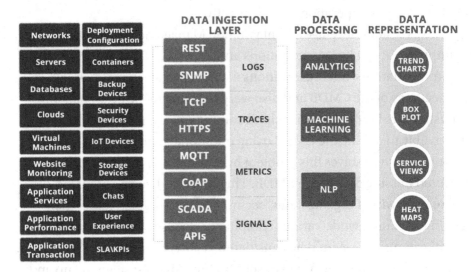

***Figure 1-2.*** *Event correlation design with AIOps*

# Data Ingestion Layer

There are many heterogeneous entities in the infrastructure, and a comprehensive monitoring landscape consists of multiple tools and solutions to monitor them. The data ingestion layer is where data from different applications, platforms, and infrastructure layers is ingested using various integration mechanisms. Typical data that is ingested is in the form of events, logs, metrics, and traces. Popular mechanisms to ingest data are Representational State Transfer (REST), Simple Network Management Protocol (SNMP), and application programming interface (API) integrations.

# Data Processing Layer

The data processing layer is the heart of an AIOps system; it is here that AI and machine learning techniques are used to process data and generate insights. Once the data is ingested into the AIOps system, the data processing layer uses machine learning and deep learning techniques to find anomalies in the data. It also uses the metric data to predict problems that may cause incidents and disrupt the business services. This layer forms the core of AIOps as far as event management is concerned.

# Data Representation Layer

The data representation layer acts as the dashboarding layer where the results of data processing layer are displayed using intuitive dashboards in various formats. The actionable data for resolution is also forwarded to external systems like ITSM so that resolver groups can act on the data and resolve the issues.

The goal here is to use the enormous amounts of data that IT systems are generating and to use AI and machine learning to make sense of that data to arrive at analytics and insights and use them to make the IT systems perform faster, better, and cheaper, and make them more resilient to failures.

AIOps helps humans to address the gap that exists in their abilities to service the needs of IT operations. It does not take away people from the operations roles but augments their capabilities to provide better on-time services leveraging AIOps.

Together humans and AI are able to deliver a level of service that both individually cannot deliver. Figure 1-3 defines how AI and human agents work in collaboration to deliver better IT operations services and which functions belong where. This humongous, ever-increasing volume of events makes it impossible for a human to analyze and then write static rules and policies. Challenges in the analysis process get cascaded to various IT services such as capacity planning, problem management, incident management, etc., who consume analysis output. AI and humans are intertwined and joined at the hip in delivering IT operations services in the AIOps model. AIOps takes the majority of time-consuming and complex tasks of data preprocessing, filtering, and analysis thereby providing key insights to the experts for making well-informed decisions.

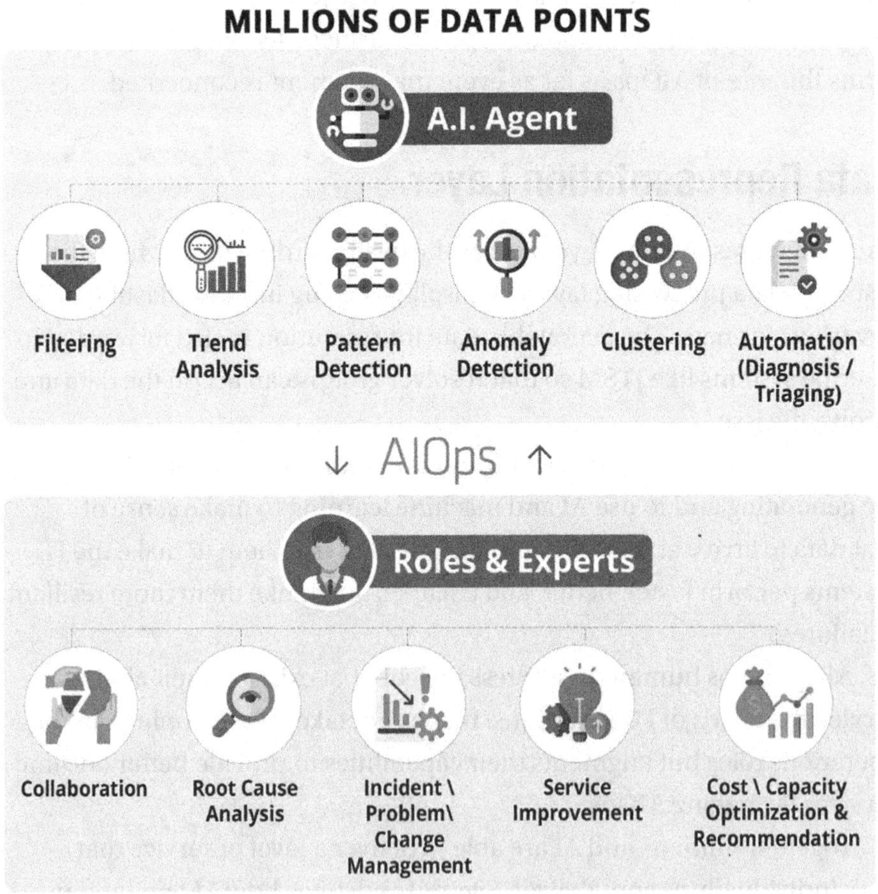

*Figure 1-3.* *AIOps-driven collaboration between AI and humans*

Through the application of AI/ML-powered data analysis and heuristics, engineers can reactively work on incidents with the aid of AIOps, which points them in the right direction and also provides them with the past data on such resolutions. AIOps is also used proactively to determine how to optimize application performance and infrastructure performance by analyzing the performance and capacity data.

Application monitoring and AIOps embedded as part of the development lifecycle would aid the development teams to proactively find availability and performance issues with either the application or the deployment infrastructure and resolve them before the application is released to production.

Adopting AIOps allows enterprises to save money by ensuring optimal utilization of capacity while at the same time avoiding downtime. If something goes wrong, the engineers are able to bring up the systems much faster than using traditional tools.

AIOps is helping to automate mundane tasks that do not require IT operators while providing contextual information for developers to improve mean time to resolution (MTTR) and customer experience. Though proactivity is core to AIOps, it applies equally to reactive situations.

Businesses are using AIOps to solve different use cases. Figure 1-4 shows the most common use cases in AIOps. Organizations start with intelligent alerting where they can do basic root-cause analysis and then move to correlation so that the root cause between various systems can be identified. As organizations move up the maturity curve, features such as anomaly detection are configured so that the operations become more proactive than reactive. Enterprises at the top of the curve have been able to deploy self-healing and automated resolution technologies so that the detect-to-correct cycle is automated completely.

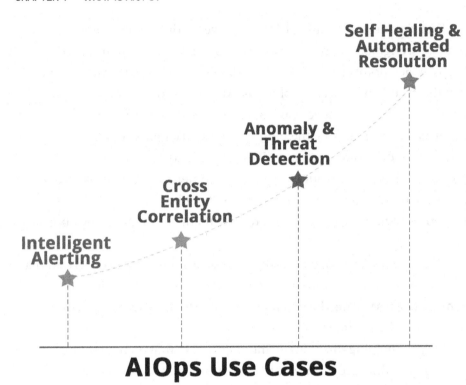

**AIOps Use Cases**

*Figure 1-4.* *AIOps key use cases*

DevOps and infrastructure operations teams have deployed many monitoring tools to get data for observability, and they are today swamped with too many events. Organizations have deployed various monitoring tools such as Nagios, Zabbix, ELK, Prometheus, the BMC stack, the Microfocus stack, SolarWinds, Zenoss, Datadog, Appdynamics, Dynatrace, etc. In addition to these tools, enterprises use cloud-native monitoring tools like Azure Monitor and AWS CloudWatch to monitor cloud-native PaaS systems. All these monitoring systems are collecting huge amounts of data from an observability perspective. Monitoring the entire stack from the network to the application is being done in many organizations. However, even with all these investments and multiple tools, organizations are struggling to get insights and actionable intelligence. The engineers are overloaded with false alerts and too many tickets to handle.

In the DevOps model, without technologies like AIOps, there are scenarios where the DevOps teams will get overwhelmed with alerts and on-call support. Bringing AIOps into the mix ensures that only actionable alerts are converted into incidents and flagged to the right teams for resolution. AIOps deployed on nonproduction systems helps to find the development and configuration issues and results in better collaboration between the development and operations teams. AIOps is instrumental in ensuring that the business services are not affected, and the right teams and resources are aligned for resolution.

Many IT teams are not well equipped to cope with the changing demands of technology. With the cloud becoming all-pervasive, the entire IT landscape is changing fast, with hybrid and cloud-native technologies being used extensively in enterprises.

Operations engineers in these situations where the transformation of the core infrastructure and application landscape is happening don't have adequate time to assess alerts and get to the root cause of the problem. In these situations, organizations are carrying a risk of unavailability and downtime.

Traditional IT monitoring and management solutions are unable to keep up with the changes in technology and depth of monitoring that are resulting in huge amounts of monitoring data being generated (see Figure 1-5). The ever-changing technology landscape means that log data and trace data are being generated at ever-increasing volumes, and it is not possible to define all rules in the monitoring systems. AIOps comes to the rescue by ingesting and analyzing all this data to make sense of it and create meaningful and relevant alerts so that the operations teams can focus on their core job of providing high availability and meeting their goals on their SLAs.

***Figure 1-5.*** *Data explosion impacting traditional IT operations*

Figure 1-6 shows the core functionality that AIOps tools can provide.

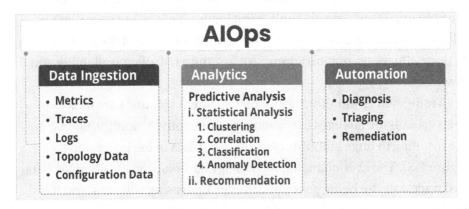

***Figure 1-6.*** *AIOps tools core functionality*

**Ingestion of data:** Data from various monitoring tools including metrics, traces, and logs is ingested, stored, and indexed for further processing. In addition, data from configuration management systems and topology data is also stored in the AIOps engine to provide correlation based on CMDB and topology relationships.

**Analytics using machine learning:** AIOps uses different types of approaches for analyzing this data to find patterns and anomalies. There are rule-based and machine learning approaches used in AIOps platforms to make sense of the ingested data. Some of the techniques that we are going to discuss in this book are statistical analysis using clustering, correlation, and classification; anomaly detection to detect anomalies in the event data; predictive analytics to find what may happen in the near future based on patterns; and topology-based and CMDB-based correlation. The idea is to convert all this event data into probable causal alerts that are the root cause for an issue so that the operations teams can focus on this and resolve the incident in a timely manner.

**Automated diagnosis and remediation:** Most AIOps tools today focus on and deliver the functionality until analysis. Automation is not part of a majority of AIOps toolsets. However, there are a few tools like DRYiCE iAutomate that apply the previous techniques to diagnosis and remediation as well, where the engine takes the probable cause as input, provides the remediation, and runs the remediation automatically. This results in automated healing and provides a complete end-to-end workflow. Let's discuss the benefits of AIOps in detail.

# AIOps Benefits

Enterprises that have deployed AIOps solutions have experienced transformational benefits. Some of them are as follows:

- *Higher availability of systems*: This is one of the key reasons and benefits of AIOps that ensures continuous services and uninterrupted business. AIOps proved to be a potential game-changer, ensuring maximum availability in today's hybrid infrastructure running containerized applications.

- *Reduction in human errors*: Due to increasing complexity and the rate of change in the infrastructure ad application landscape, the majority of the outages happened due to human errors. This is another lever for AIOps system adoption because AIOps automates most of the repetitive and mundane tasks.

- *Better SLA compliance on mean time to repair*: This is the target goal of any IT operations and a genuine expectation from the business. AIOps system integration with ITSM functions makes it feasible by uncovering useful insights, finding patterns of issues, and enabling collaboration with automation solutions to resolve them quickly. All this means that the mean time to repair is reduced and helps IT operations teams to not only meet but exceed the current SLAs.

- *Better automated detection of incidents*: This is another key benefit of AIOps. An AIOps system eliminates a lot of waste by reducing the noise that gets created due to the creation of false-positive incidents. An AIOps system leads to the thorough analysis of events to qualify for the incident creation with appropriate severity. This saves IT operations teams' time, which is wasted when chasing false positives.

- *Prediction and prevention of outages*: AIOps leads to proactive operations and an important KPI to measure the operations performance. The AIOps system generates intelligent recommendations that help IT operations to meet this objective.

- *Cost optimization*: IT is still being considered as a cost to many organizations. A mature AIOps system drastically brings down operational costs. By offloading work to algorithms and freeing up the human resources to spend time and energy on value-adding items, organizations are better able to utilize their precious human resources.

- *Better visibility into the environment*: AIOps not only enables IT operations to identify areas of improvement but also enables businesses to uncover new opportunities or take strategic decisions. As AIOps systems touch all IT functions, they are best suited to filter out the noise and provide relevant visibility of the IT estate being managed to stakeholders.

- *Reduced risk of operations*: Risk management is one crucial domain in IT operations, but an AIOps system taking charge of automated execution of tasks, reducing human errors, and enhanced analytics using AI-powered tools greatly reduces the operations RISK irrespective of whether it is related to security, disaster recovery (DR), or day-today operational tasks of incident management, change management, and problem management.

- *Automation benefits*: Automation is a journey, but it often fails or does not deliver expected results when it works in silos. An AIOps system, on the other hand, enables the integration of core IT functions by providing end-to-end automation services.

- *Higher maturity of IT operations*: AIOps' continuous feedback provides visibility into gaps and challenges in processes, tools, and infrastructure. This leads IT operations from a reactive state to a mature proactive state.

- *Better visibility, governance, and control*: Organizations often implement various event management and reporting tools for operational governance and control but often fail due to the dynamic nature of the infrastructure and the inability of the operations team to keep the systems updated. AIOps system, on the other hand, can automatically detect and absorb such changes using algorithms and deliver the required visibility for governance and control.

- *Easier to move to SRE, the DevOps model*: The AIOps system brings automation and maturity in IT processes and tools, thereby enabling the operations team to adopt SRE and a DevOps model.

- *More efficient use of infrastructure capacity*: An AIOps system provides much more efficient and granular visibility into capacity utilization, enabling the capacity manager to perform demand-forecast and cost-benefit analysis in a much better and faster way.

- *Faster delivery of new services*: An AIOps system eliminates wastes, upskills the operation team, and brings maturity in processes and tools. This enables IT teams to support new initiatives and services.

# Summary

In this chapter, we covered the challenges being faced by operations teams and how AIOps helps organizations overcome these challenges. We discovered AIOps and its various components and the capabilities that each of these components provides. We also listed the benefits organizations can expect when they deploy AIOps. In the next chapter, we will explore the AIOps architecture and methodology.

# CHAPTER 2

# AIOps Architecture and Methodology

In this chapter, you will learn about technologies and components that are in the AIOps architecture along with its implementation methodology and challenges. The chapter will explore the AIOps key features of the Observability, Engage, and Act phases and the role of machine learning in IT operations.

## AIOps Architecture

AIOps systems primarily consist of three core services of IT operations, which are enterprise monitoring, IT service management, and automation. The AIOps architecture provides technologies and methods for seamless integration between these three services and delivers a complete AIOps system. Figure 2-1 defines the AIOps platform and its applicability across various processes and functions in the three core services defined in the IT operations value chain as defined by Gartner. Let's delve deeper into this and understand the AIOps architecture.

© Navin Sabharwal and Gaurav Bhardwaj 2022
N. Sabharwal and G. Bhardwaj, *Hands-on AIOps*,
https://doi.org/10.1007/978-1-4842-8267-0_2

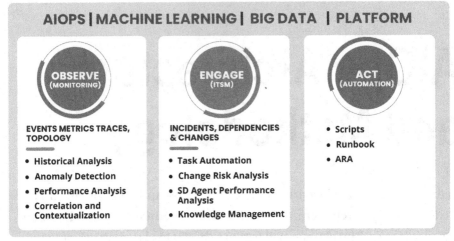

**Figure 2-1.** *AIOps architecture*

# The Core Platform

The AIOps system is intended to ingest millions of data points that get generated at a rapid pace and need to be analyzed quickly along with historical data to deliver value. Big Data technologies coupled with machine learning algorithms provide the solution and form the core of the AIOps system.

# Big Data

The AIOps platform ingests data from multiple sources, and thus the platform needs to be built on Big Data technologies. Traditional systems have been built on relational systems; however, when it comes to AIOps, there are multiple data sources and a huge amount of data to be processed to arrive at meaningful insights. Big Data is defined as the type of data that is characterized by the five *V*s, which are volume, velocity, variety, veracity and value, as shown in Figure 2-2.

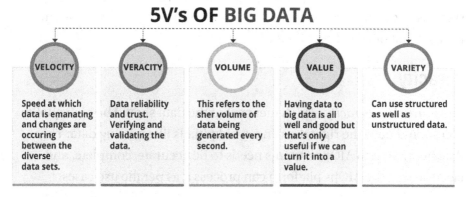

**Figure 2-2.**  *Big Data definition*

## Volume

*Volume* is the core characteristic of Big Data. The amount of data in Big Data systems is much larger than the data handled by RDBMSs. As AIOps integrates data from various systems into a data warehouse, the volume of data becomes unmanageable without Big Data technology platforms at the core.

## Velocity

The second *V* in Big Data technologies is *velocity*, which describes the speed of data that is sent for processing. Since AIOps deals with event data that has to be processed quickly and in real time, velocity is an important parameter in the data that is being processed with AIOps. The data needs to be processed at near real-time intervals so that the response from human or machine agents is immediate. Data from multiple sources is sent to the AIOps systems at high velocity, and this needs the appropriate platform architecture to process this data at scale and with speed.

## Variety

The third V of big data is *variety*. In AIOps there is a variety of data that needs processing. There are multiple monitoring and management systems and data such as events, logs, metrics, tickets, etc., with varying formats that need to be processed in AIOps.

## Veracity

The fourth V is *veracity*, which means that the data should be high quality. You should have the right data and avoid pitfalls like missing data. The data being sent to AIOps systems needs to be accurate, complete, and clean so that the AIOps platform can process it as per the use cases.

## Value

*Value* means that the data should have business value. The AIOps data is highly valuable as it translates to higher availability and visibility, and it forms the backbone of automation to reduce costs and provide better services to customers.

# Machine Learning

Apart from Big Data, a core component of AIOps is machine learning. Artificial intelligence and machine learning technologies are at the core of AIOps. Traditional systems have been used for monitoring and event correlation; however, they were rule-based and did not use machine learning technologies to efficiently and effectively derive insights and provide features and advanced use cases that are possible by leveraging machine learning technologies.

AIOps platforms leverage the power of machine learning to analyze data being fed from various systems and to detect relationships between monitored entities and events in order to detect patterns and anomalies.

We will be discussing the machine learning aspects of AIOps in detail in the upcoming chapters.

This data is then used to provide insights and analytics and arrive at root-cause alerts. AIOps platforms combine CMDBs, rule-based correlation, and unsupervised and supervised machine learning to achieve the end objective of finding out the root cause, and they attempt to provide predictive insights to find hidden issues, which may surface later. The core themes are the higher availability of systems, better insights and governance, and higher customer satisfaction scores.

Let's understand how AIOps improves the IT operations.

# The Three Key Areas in AIOps

AIOps cuts across three key areas in IT operations, which are Observe, Engage, and Act.

Traditionally, some aspects of the Observe area were under Monitoring, and now with end-to-end visibility being the focus, it has matured to "observability."

The second key area is Engage, which is part of the value stream and is related to IT service management and the functions of IT service management such as service desk, command center, and resolution groups as well as the ITSM processes such as incident management, change management, problem management, configuration management, capacity planning, and continual service improvement.

The third area is Act, which defines the technical function where technical teams resolve incidents, complete service requests, and orchestrate changes in the IT systems.

We will now delve deeper into each of these areas and see how AIOps impacts them.

# Observe

Unlike traditional monitoring and event management tools, Observability uses machine learning–driven functions and ensures there are no blind spots or gaps left while catering to the enterprise monitoring needs of organizations irrespective of whether monolithic applications are running on traditional physical or virtual infrastructure or modern applications running on cloud-native or microservices architectures. Primarily four processes get performed in this stage, as shown in Figure 2-3.

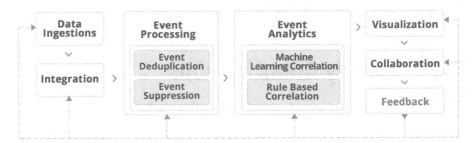

***Figure 2-3.*** *Observability using AIOps*

# Data Ingestion

Data ingestion in AIOps is the first important step, and all monitoring and management data is ingested into the AIOps system so that all the data is available to the system for analysis. At times while implementing AIOps projects it is observed that the basic monitoring data is not in place while the organization is insisting on going ahead with AIOps; in such situations a fundamental discussion of how AIOps works and how machine learning algorithms are completely dependent on data comes to the rescue. Setting the right monitoring pieces can be spun off as a separate project under the program while AIOps continues to integrate and move forward with the plan of integrating its data sources. When the entire data is available, the algorithms are trained and tweaked to reflect the new data.

For event management, the following data is required:

- *Events*: These are the events generated from various sources including operating systems, network devices, cloud computing platforms, applications, databases, and middleware platforms. All these platforms generate events, and they are captured through monitoring tools and then forwarded to the AIOps systems.

- *Metrics*: These are performance metrics, which include infrastructure metrics such as CPU utilization, memory, and disk performance parameters, network utilization, and response time metrics. These also include application metrics such as response time of an application, page load times, the completion time of queries, etc. The metrics are collected in a periodic manner, say every five minutes, and the data is used for understanding the behavior of the system over a period of time. This data is also termed *performance metrics*.

- *Logs*: Many systems maintain log files and provide data in logs. The log collection can be configured and tuned to log certain types of information. These log files are sent to AIOps systems to find patterns.

- *Traces*: The applications use tracing mechanisms to provide information on a complete application transaction right from the users' browser to the backend servers; these are logged using various mechanisms including widely used formats like OpenTracing. The trace data provides information on the end-to-end transaction, and it provides the path and time that each step of the transaction took. Any errors or performance issues in any component can be diagnosed by using the trace data.

Apart from the real-time data mentioned, AIOps tools need discovery and configuration data as well so that topology and relationship-based correlations can effectively work. This data ingestion may be done periodically rather than on a real-time basis.

## Integration

For data ingestion to happen, there needs to be integrations available in AIOps platforms. AIOps platforms should support both push and pull integrations. In a push model, the monitoring tools or forwarders can forward the data to the AIOps engine. In a pull model, the AIOps tools have the capability to pull the data from various monitoring systems.

The other aspect of integration is real time and periodic data upload; the event, metric logs, and trace data is integrated on a real-time basis since the operations teams need to take immediate actions. The discovery and CMDB data can be integrated with the AIOps system periodically, say, every day and weekend as a batch job or just after the synchronization of the CMDB jobs. This helps in keeping the AIOps data up-to-date on the topology and relationships. However, in cloud computing and software-defined environments, the discovery of CMDB data is real time, and the integration is done on a real-time basis during provisioning and deprovisioning using infrastructure as code.

The adapters for integration to various monitoring and management systems are part of the AIOps system. These need to be configured to bring data from different systems into a single Big Data repository for data analytics and inference.

## Event Suppression

Using the integrations and data ingestion, all the data from various monitoring and management tools is ingested into the AIOps solution. This voluminous data needs to be cleaned to reduce the noise. The first step is event suppression, where unwanted alerts are suppressed or eliminated from the system. Care should be taken that any data that is relevant for further processing or can provide any information to the AIOps engine is not discarded during suppression. An example of suppression may be the information events in event and log sources. These events are only for informational purposes and do not indicate an underlying problem in the system. Another type of data that may be there in ingested log files could be the successful executions; this data is again for information and not relevant for finding any issues with the system. In other systems, events that are warnings may also be discarded during the event suppression phase; this decision needs to be taken by the domain experts.

Without event suppression, the AIOps system will be overloaded with data, which is not needed for further processing and analytics.

# Event Deduplication

Once events are suppressed, the next step is event deduplication. Deduplication is an important step in the processing of AIOps data. In this step, duplicate events are clubbed together and deduplicated. As an example, if a system is down, the monitoring tool may send in this data every one minute; the data shows the same information but with a different timestamp. The AIOps system takes this data and increments the counter of the original event and then updates the time stamp to reflect when the last event for this system and this particular event occurred. This gives the required information to the AIOps system as well as the engineers on what systems are down from what time and what is the timestamp of the last data point. Deduplication ensures that the event console is not cluttered with multiple events in the console and relevant information is available.

Deduplication preserves the information that is sent by successive events that remain in the system and is used for processing by the AIOps system; however, rather than showing it multiple times, the information is aggregated in the console and database.

Without event deduplication, the event console will become cluttered with many events showing the same event again and again.

## Rule-Based Correlation

AIOps systems use machine learning technologies to analyze the data. Traditional systems used rule-based correlation to decipher and analyze the data. In AIOps systems, rule-based correlation still plays an important role, and there may be policies in an organization that need implementation through rule-based configuration rather than using probabilistic machine learning models. An example of this could be a rule that increases the severity of an event based on the type of a system being development or production. This is achieved by looking up the CMDB to find the type of the device and then applying a rule or policy to upgrade or downgrade the severity of the event based on the system's categorization. There are other important rules that are required in AIOps systems like "maintenance window," where the alerts are suppressed for the period of any maintenance or patching activity. This reduces noise in the system and prevents the event console from showing alerts for systems that have been rebooted or shut down for maintenance activities.

Rule-based correlation also has a subtype called *topology-based correlation*. Topology and relationships between systems are used both in rule-based correlation as well as in machine learning–based correlation. In rule-based systems the topology of the system and their relationship is used to suppress or correlate events. For example, if a switch is down, then all servers or infrastructure beyond the switch is unreachable from the monitoring systems. The topology-based correlation will flag these events, correlate all infrastructure events with the switch-down event, and flag the switch as a probable root cause for all these events.

## Machine Learning–Based Correlation

The machine learning–based correlation that is provided by AIOps tools is what makes it AIOps. The other features that we discussed were available with traditional platforms, but machine learning–based capabilities are the ones that differentiate AIOps products from other event management and event correlation engines. Figure 2-4 shows various types of correlation that are performed by the AIOps engine.

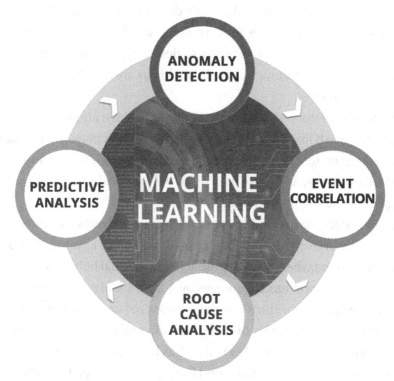

*Figure 2-4.* *AIOps-driven correlation engine*

In this section, we will take a deeper look at each of these types, beginning with anomaly detection.

# Anomaly Detection

Anomaly detection is the process where you can identify unexpected events or rate events in the data. Anomaly detection is also termed *outlier detection* since it is about detecting outliers or rare events.

The outlier or anomaly events differ from the normal events. Anomaly detection algorithms try to identify these events from normal events and flag them as anomalies. Typically, events that are anomalies may point to some issues with the system. Anomaly detection techniques have been used in various use cases such as fraud detection in credit card and banking transactions, security systems to detect cybersecurity attacks, etc. With AIOps the same techniques and algorithms are now being used on the IT operations data.

Anomalies are not just about rare or outlier events; anomalies are also detected in AIOps systems for metric data such as network or system utilization parameters. In metric data, the anomalies are the bursts in utilization or activity, and these may point to some underlying causes that are flagged by the AIOps systems.

Anomaly detection techniques use unsupervised and supervised as well as semisupervised machine learning to flag events as anomalies and flag the metric data when there are breaches to normal behavior.

Anomaly detection has advantages over traditional rule-based systems. Anomaly detection algorithms can detect seasonal variations in data and only flag anomalous behavior of the system after taking into consideration the seasonality of variations. Metric data has high seasonality since application load and jobs running on IT infrastructure typically follow a time-of-day season. Some of the jobs that are run monthly also increase the utilization of systems but are not anomalies.

Anomaly detection can be done using a variety of available algorithms in AIOps systems. The AIOps teams can use these algorithms to fine-tune the implementation based on the type of data and in the environment.

Thus, AIOps systems using anomaly detection are better suited to reduce the noise in the event data or metric data by flagging the right events based on the seasonality of data on one hand and by finding anomalous patterns that may get missed in rule-based systems. Together these features help the operations teams by giving them insights into what is happening in their environment so that they can take reactive and proactive steps to remediate problems or prevent problems from occurring in the environment.

## Event Correlation

Modern digital applications are all interconnected. Even traditional applications have been developed using distributed architecture where web servers, application servers, and database servers work in tandem to complete the application functionality. The infrastructure itself is distributed in network topologies with routers, switches, and firewalls routing the traffic from users of different locations to the main data center where the application is hosted.

The distributed application and infrastructure are monitored using multiplicity of tools. Thus, an environment will have alerts coming in from the network monitoring tools, server monitoring, database, and platform monitoring, and the application itself will be logging events and traces. All this data needs correlation so that the noise can be eliminated, and the causal event (the event causing the problem) is flagged and identified automatically. In the scenario where AIOps tools are not deployed, this activity is done by different subject-matter teams coming together over a call that is termed the *operations bridge* or *critical incident bridge* and analyzing all the systems and data to collectively identify what could be the potential source of the problem in a distributed system. You can imagine the complexity and the time taken to go through all the data and arrive at a conclusion.

Event correlation also takes feeds from configuration management and change management systems, thus correlating these changes in systems to the events getting generated by monitoring tools. This helps in root-cause analysis as many incidents and problems arise after making a configuration change or patching an existing system. Change and configuration management data needs to be made available to the AIOps engine to correlate with event and performance data coming in from the monitoring systems. In fact, the first thing subject-matter experts look for is any changes that may have been made to the system recently that would have caused the incident.

Machine learning–based event correlation helps solve this problem by automatically grouping related alerts by correlating across various parameters so that the resolution group gets all the information at a single place. Event correlation is done using the topology and relationship information that is available in discovery and CMDB systems; it also uses timestamps and historical data to group events and provides insights to operations teams.

With time and enough supervised learning data, the event correlation engine becomes more accurate. The event correlation engine feeds in the data to "root-cause analysis." We will be covering root-cause analysis in greater detail in the next section. Without event correlation, it is not possible to do root-cause analysis or predictive analytics. Thus, event correlation is the first step to move forward with root-cause analysis and predictive analytics.

## Root-Cause Analysis

Root-cause analysis is the most important module of AIOps; this is where the most value is realized by operations. We have already seen in event correlation how modern distributed applications and infrastructure are spread out and events are generated from various monitoring and management systems and how this impacts operations' ability to go through this large set of data and try to find the root cause of the problem manually. With such complexity coming into operations, it is becoming impossible to do root-cause analysis without the aid of event correlation systems. Whether an organization uses a rule-based approach or an AIOps-based approach, without leveraging technology, it is impossible to do root-cause analysis and remediation of issues and meet the service levels that are agreed with the business.

Identifying the root cause manually or with event correlation and automated analysis involves multiple teams from different IT domains coming together and analyzing the situation to arrive at a conclusion on what could be the problem. This also involves the need for collaboration and tools for collaboration so that different stakeholders are on a common platform and can effectively do "root-cause analysis." Root-cause analysis with AIOps looks at all the data being ingested into the AIOps system and provides insights to the teams for faster and better root-cause identification. Since machine learning technologies are probabilistic in nature, root-cause analysis in AIOps parlance is also referred to as *probable cause analysis*; thus, it may throw multiple probable causes for a root cause with an attached confidence score. The higher the confidence score assigned to an event, the higher the probability that the AIOps engine has assigned to the event to be the root cause. Based on the probable cause, the operations teams can do a deep dive and arrive at the final root cause and mark it as such in the system.

Root-cause analysis leverages anomaly detection and event correlation techniques as well as supervised learning feedback to arrive at the root cause. Root-cause analysis leverages both supervised and unsupervised techniques to arrive at this result.

Since the IT data is vast and dependent on the environment, the feedback loop is an important aspect of root-cause analysis. There are vast amounts of tribal knowledge in the minds of the operations teams that may be undocumented and known only in the form of informal knowledge. When the operations teams collaborate and mark the root cause from various probable causes that the system has generated, the AIOps systems learns from the human operators and updates its model. Thus, the system can store the actions taken by operations teams and can recall the root cause from previous incidents. We will cover this in greater detail in the feedback section; however, it is important to understand that root-cause analysis has a dependency on human feedback, and without this, the root-cause analysis accuracy may hit a ceiling.

The benefits of automated root-cause analysis are many. There is a marked improvement in the mean time to respond and mean time to resolution since the system makes the job of human operators easier by flagging the probable cause and eliminating noise from the system. Integrated with knowledge management and feedback loop from operators, root-cause analysis creates a robust and highly accurate system as its usage gets expanded over time.

There are certain limitations of the root-cause analysis using machine learning technologies. Since this is probability based, there is no guarantee that the root cause identified is the actual root cause. The other limitation of root-cause analysis is that unlike anomaly detection and predictive analytics, which can be done using data alone, root cause is dependent on human feedback. If there is no participation from the operations teams, the root-cause analysis will remain at low levels of accuracy. Thus, the human element, feedback, and training of the AIOps engines are important parameters and limitations as well. Trying to do root-cause analysis with only unsupervised learning will not be effective as it will only be able to flag anomalies, but whether the anomaly is actually causing a system degradation or an incident may not be accurately flagged by the AIOps engine. Another limitation of root-cause analysis is the nature of incidents in the IT domain; an incident may be unique and have a unique combination of events that are generated during that incident, and that incident may not have happened in the past, so there is no precedence or data in the AIOps engine that it can use to arrive at a conclusion. Thus, novel or new incidents with new events previously unseen are a challenge for current AIOps systems.

Thus, we cannot expect the current generation of AIOps systems to accurately pinpoint root cause without training by subject-matter experts and human operators. There are scenarios where the customers expect the AIOps engine to be a magic wand and automatically start finding the root cause and remediation of problems; however, deep learning and machine learning systems are dependent on labeled data and training, and without this training, the system is incapable of providing accurate results.

Since root-cause analysis is the most important and complex lever in the entire AIOps stack, it is important to pay utmost attention to its implementation and continued operations effectiveness. As a future direction, AIOps tools can use multiple algorithms and an ensemble of algorithms to do root-cause analysis and provide better accuracy even with limited training data.

Root-cause analysis feeds into automation; without having this process step trained and generating accurate results, end-to-end automation and remediation are not possible. Once the root cause is identified, it feeds into the automation engine to resolve problems automatically and catapult the organization into the highest levels of maturity with autohealing.

## Predictive Analysis

Predictive analysis brings in the predictive element to IT operations. This is an area where the customers have always expected predictive capabilities from IT operations systems but have not been able to achieve them. AIOps brings predictive analytics capabilities to IT operations and fulfills this unmet demand from the operations teams.

As the name suggests, predictive analytics means the ability to predict things in advance based on the data provided to AIOps systems. There are use cases where predictive analytics has a role to play in the IT operations space, so let's take a look at some of them.

One important application of predictive analytics is in the area of performance management and capacity planning. With metric data being available to AIOps systems, it is possible for the AIOps engine to predict what the future utilization of these systems will be. The data around users accessing an application and the associated utilization of the system can be used to make predictions based on scenarios of how many users will hit the application and how much infrastructure will be needed to support these users. Regression techniques can be used to consider the current performance and workload of a system and predict the future utilization of the infrastructure. Being able to predict in advance the utilization helps IT operations teams to plan better for infrastructure capacity and instantiate new virtual machines or cloud instances to cater to the predicted demand. In microservices-based applications, new pods are spun automatically to cater to increased demand on infrastructure.

Predictive analytics uses regression techniques, which can take into consideration the seasonality of the data and provide accurate results. As an example, a backup or data processing job at the end or beginning of a month may be causing performance issues and errors in an application. Leveraging predictive analytics techniques can ensure that the AIOps system is able to forecast the utilization and the operations teams can take appropriate actions to increase the capacity during that period by spinning extra instances or by vertically scaling up the capacity so that there are no performance issues and incidents are avoided.

Another use case based on metrics can be around trend finding, where the AIOps engine is able to spot a trend and an associated event at the end of a trend. Based on this association, it can predict in advance things like a system failure. An example of this would be a memory leak issue in an application that causes the memory utilization of a machine to keep increasing and thus following a trend. After consuming all the available memory, the system starts to consume the disk space, and the memory is paged to the physical disk. After a while, the application starts to slow down and eventually crashes, causing a set of events. This pattern can be detected by predictive analytics engine, and on observing, a trend the AIOps system can warn the operations teams of impending failure.

Another example on the previous lines can be of a faulty database connection code where the database connections are not released and after a while choke up the entire database and the application starts to get connection failure alerts. The metric around database connections when plotted will form an increasing trend and can be deciphered by the AIOps systems to forewarn the operations teams.

Like other elements in the AIOps space, predictive analytics is also based on probability and thus may not be 100 percent accurate. Predictive analytics in the IT operations space looks at events and metrics to predict what could be a probable outcome and then alerts the operations teams.

The customers sometimes think of AIOps tools as a magic wand and expect them to predict all kinds of failures and prevent them. This is impossible with the current state of the technology as all failures are not predictable. Only events that have underlying patterns decipherable through trends or a sequence of events prior to the failure can be deciphered by the AIOps engine. We have to be aware of the limitations of the system and configure them to their best capabilities and not expect magic. There are lots of failures that are unpredictable in nature and happen at random. Devices and systems fail randomly without any forewarning, and predicting their failure accurately is not possible with the current state of the technology.

Predictive analytics can be simple based on a single variable, or it can work on multiple variables and the correlation between them to arrive at a prediction. Predictive analytics systems can take care of seasonality of data in their models to arrive at accurate predictions with a high level of accuracy.

Predictive analytics results in proactivity in operations and a higher availability of the system since the problem is remediated before it is able to impact the availability or response time of an application.

# Visualization

Visualizations are important parameters in the AIOps tools. There are various types of views and dashboards that are needed from an operations point of view in AIOps tools.

The foremost visualization is the event console. The AIOps tools need to have an intuitive and easy-to-use event console. The event console is a grid view that has all the alerts that need action or analysis from the operations teams.

The following is important information that is available in the event console:

- Event ID/alert ID

- Description

- First occurrence

- Last occurrence

- The number of times the event has occurred

- Any associated incidents with the alert

- Severity of the event

- Whether it is a probable cause or not

- Status whether it is open or cleared

- History event/alert with associated actions and state changes

The event console typically color codes the events per their severity and their tagging of being probable cause or not. Events that are correlated together are shown together in a consolidated console that enables the operations teams to look at all correlated events associated with a probable cause in a single console and deliberate on the root cause.

Beyond the event console, the AIOps console has other dashboards that provide aggregated and consolidated information, as shown here:

- Event trends, patterns; graphical plot of event trend

- Top events across the environment

- Top applications or infrastructure elements causing events

- Information on event flood

- Performance data plots for metrics

- CMDB views/topology views

- Historical data around events, alerts, and performance metrics

Apart from the previous views, the AIOps engine may also provide views and information on the performance of the AIOps engine itself.

Visualization in AIOps engine thus comprises the Event console, dashboards for real-time data, and reports for historical analysis of data that drives the collaboration process and is going to be discussed next.

## Collaboration

IT operations teams collaborate to find root cause and to deliberate on ways of solving the problem. The Command Center function in IT operations is where the collaboration between various teams happens over a bridge. A bridge is an online real-time call over communication channels like Microsoft Teams and Phone where multiple teams interact and collaborate to look at the events and incidents and try to analyze and find the root cause of the issue at hand. Priority 1 incidents that are impacting an application or infrastructure mandate the opening of a P1 bridge where the required stakeholders from different technical domains collaborate.

In AIOps, the same process runs; however, there are a few differences. More and more teams are leveraging the built-in ChatOps features of AIOps tools where different team members can converse as well as run scripts to diagnose and resolve the problem.

Another change in AIOps is that rather than looking at different consoles and events, the entire team has access to the AIOps event console where consolidated and correlated events along with probable cause that are flagged by the AIOps system are available for the teams.

The views on topology and relationships between the affected or systems under investigation is also available from the AIOps console; thus, the teams don't have to go to various systems to get the complete picture.

This results in speeding up the entire process of root-cause analysis, problem identification, and resolution of the problem.

Another important aspect of collaboration in AIOps is that this is no longer a collaboration between people only. The artificial intelligence system is a party to the entire collaboration and is storing the information in its records to be used as a learning tool and to be used in future incidents involving the same set of events or same probable cause. AIOps tools can bring up the past records and help the operations teams in referring to the accumulated knowledge from past collaborative analysis that was done. Thus, historical knowledge is not lost but accumulated for usage in the future.

## Feedback

Feedback is the last step in our Observe process, but it's perhaps the most important step. As you learned in earlier sections, root-cause or probable cause analysis is one of the most important use cases in AIOps, and the foundation of root-cause analysis is the continuous feedback on its accuracy and confidence scores provided by the operations teams. Every root cause identified by the AIOps engine is analyzed, and feedback is provided in the system by the operations teams. Thus, an incorrect root cause provided by the AIOps engine is marked as incorrect, and the correct ones are marked as correct. This data feed helps the AI engine to understand the environment and improve on its model. This data is the labeled data that is required for training the supervised learning system in AIOps. Once there is enough data with the AIOps engine on what events are root cause and which events are not root cause, it is able to better analyze and interpret the next set of events based on this learning. Thus, feedback is what powers continuous learning of the system. This enables the AI system to learn and improve its accuracy and confidence scores and achieve a level of accuracy where the data can be then used to initiate automation.

Typically, after a few months of running the AIOps system and providing the correct feedback, the system's accuracy and confidence scores reach a level where the automation can be initiated from the AIOps engine for high-confidence probable-cause alerts so that the entire process from detecting a problem to taking a correction action through automation is achieved automatically without human intervention. This concludes the Observe stage of AIOps system. Let's now move to the other core function under AIOPs, which is Engage.

# Engage

The Engage area is related to ITSM and its functions. It is an important piece in the AIOps space as it primarily deals with the processes and their execution by various functions and the metrics around process and people. The Engage piece deals with the service management data and hence is a repository of all the actions taking place in important ITSM functions like incident management, problem management, change management, configuration management, service level agreements, availability, and capacity management. Figure 2-5 illustrates this.

**Figure 2-5.** *AIOps-driven IT service management*

Continual service improvement is an important lifecycle stage in ITSM, and that's where most of the analytics is performed in AIOps. In Observe, the primary data includes events, metrics, logs, and traces, but here the primary data is around the activities being done in various processes. Workflows in Observe are more machine to machine; here the workflows involve the human element.

The data in Observe is mostly real time, but in Engage it is a mix of real time as well as on-demand analytics.

Let's do a deep dive on this and understand its elements and stages.

## Incident Creation

The Engage phase starts with the Observe phase creating an incident in the ITSM system. After probable cause analysis creates a qualified alert, the alert is sent to the ITSM system for creation of an incident. Incident creation needs various fields to be populated in the ITSM system so that the information is complete and helps the resolution teams in resolving the incident. The AIOps Observe tool and the ITSM tool are integrated to automatically create incidents in the ITSM system and have the fields in ITSM autopopulated from the information available in the Observe module. This includes the description of the alert and other related information as defined earlier in the Observe section.

If an alert is cleared from the Observe module, the AIOps engine keeps updating the incident in ITSM automatically and marks it as cleared so that the incident can be closed. If the alert gets new events that are triggered, it keeps updating the incident in the ITSM module with new information that alerts the operations teams.

There are scenarios where the events in the Observe console do not get autocleared if a problem is resolved. In those scenarios, there is a two-way integration where the ITSM system clears the alarm when an incident is closed so that the Event console reflects the accurate state of systems being monitored.

# Task Assignment

In traditional systems, tasks are assigned to engineers by track leads based on the availability of the resources and the skill required to deliver a particular task. In modern AIOps-based systems, the task assignment is done through automation defined in the ITSM system or outside of it. The task assignment engine takes into consideration the availability of the resource in a particular shift, their skill level, the technology required to solve a particular task or incident, and the workload that is already with the resource. Based on these parameters, the ticket is assigned to an individual to work upon and update the progress until closure.

Task assignment is done using rule-based systems rather than machine learning systems since it matches the skills to a task and matches the experience level of a resource along with the workload or availability, and these lookups can be done using rule-based systems rather than leveraging machine learning technologies.

However, natural language processing and text extraction–based systems can be used to extract the information from the incident and map it probabilistically to the right skill and thus aid the task assignment engine. The machine learning capabilities using text extraction help in automatic mapping of the task to the right skill rather than using a regular expression–based approach to look for keywords. Using or not using machine learning for this is entirely dependent on the scale, size, and complexity of the environment. For smaller environments, rule-based systems would work perfectly well, and leveraging machine learning here may not be needed. However, larger and more complex operations would require this capability to run efficient operations.

## Task Analytics

Tasks assigned to individuals need to be analyzed; thus, each task that is in the system is generating data. Statistical analysis of the tasks in the system is used to provide insights into how the process and people are performing. Tasks can be analyzed for volumetric data as well as for efficiency data in terms of time taken at each step. Analyzing the tasks gives important insights to run Six Sigma or Lean projects in the organization.

This is also used for assessing the accuracy of the assignment engine to see if the tasks are correctly assigned. If the tasks are not correctly assigned, the task will keep hopping between different teams, and this may indicate a problem with the assignment engine.

## Agent Analytics

Similar to task analytics, the other important lever in ITSM is the agents or resources working on these tasks. Agent analytics analyzes the performance of human as well automated agents on parameters such as accuracy, time taken to resolve issues, individual performance, and performance as compared to baselines. This can flag any issues with skills or availability of resources. This data is also useful to analyze if the assignment engine is assigning tasks correctly.

# Change Analytics

Changes including patches, updates, upgrades, configuration changes, and release of new software into production are potential sources of incidents. What was working before may stop working after making a change. Thus, it is important to analyze changes that are happening in the infrastructure and application environment.

Change analytics includes areas where the impact of change can be assessed by using topology and configuration information. Change analytics also includes probabilistic analytics of risk to infrastructure and platforms because of changes. This may involve analyzing the data around topology, relationships between various components, the size and complexity of the change involved, and the associated historical data with these changes to arrive at a risk score for a particular change. The feedback score from technical evaluators and business approvers of change is also an important input to analyze the change and plan for its execution, keeping in mind the risks that it carries.

# Process Analytics

We talked about analyzing key basic processes in ITSM including incident management and change management. However, all processes in ITSM need analytics particularly around the KPIs that are defined for each process.

As an example, change management has associated KPIs for changes implemented within a certain time, changes that caused outages and incidents, etc. Similarly, incident management has KPIs around response time and resolution time of an incident along with other process KPIs such as time taken to identify root cause, etc.

Service level management processes have KPIs around the SLAs that may be related to the response and resolution of priority-based problems. For example, all P1 incidents should be responded to within 5 minutes and resolved within 30 minutes with an SLA of 90 percent over a monthly cycle. This means that 90 percent of the P1 incidents should be responded to and resolved within the time defined, and this calculation is done on a monthly basis and reset at the beginning of each month.

All this process data is fed into the AIOps analytics engine to do statistical analysis of this data and analyze it for process improvement purposes. You can also use AIOps machine learning techniques like regression here to predict the future metrics based on the historical data; the regression techniques will take into consideration the seasonality variations and the data from the past to arrive at the future predicted values.

This data helps to better plan the resources and also feeds into process improvement initiatives.

## Visualization

Since most of this data in ITSM systems is about people, process, and technology aspects, it is important that we have the right level of visualization and dashboarding technology available to make sense of this data and walk the path of continual service improvement.

There are various stakeholders who need access to this data, and their requirements are different; thus, the visualization layer needs to have role-based access and role-based views to facilitate the operations team.

There are service delivery managers, incident managers, command center leads, process analysts consultants, and owners. You also have the service level manager and change and configuration managers who are responsible for the SLAs with customers and responsible for maintaining the CMDB, respectively. All these roles require the right level of insights and visualization into the relevant data to be able to manage their respective processes.

Visualization is also needed for business owners and application owners, and in the case of outsourcing engagements, there are views needed by the customer and service provider.

Right dashboarding and visualization tooling in AIOps is essential to get all the required data and insights including insights generated by machine learning algorithms to run operations with higher efficiency and maturity.

# Collaboration

Just like Observe, collaboration is important in the Engage phase as well. In the Observe phase, the collaboration between teams happens on a bridge or using ChatOps to find the root cause of a problem. In the Engage phase, collaboration happens between various stakeholders to resolve the problem. Thus, various stakeholders collaborate on the tickets to bring the problem to closure. Unlike Observe, where multiple teams have to come together to analyze the issues, here it is a limited set of people sometimes restricted to a particular technology domain; at times it is only one individual who is working on the ticket to resolve the issue.

Collaboration also happens in the service request and change execution task; however, most of it is orchestrated through a rule-based system where completion of each task is sequentially assigned to the person who needs to do the job. There is a greater degree of collaboration required in change management as multiple teams may be involved in executing a complex and big change; the change management process manages these through a rule-based approach where the required stakeholders are brought together by the system at various stages of a change.

If the person who is responsible for executing a task is not able to finish it within the time period assigned, the system assigns or engages a higher skilled resource to help and complete the task in time. This is all done using a rule-based engine that keeps a tab on the time to complete a task and escalates it after expiration of a set period of time.

Collaboration also happens in these processes on the visualization or dashboarding layer where different stakeholders can collaboratively look at the data, analyze it, and make decisions that require inputs from multiple teams or stakeholders.

Though largely rule based, there are aspects of ChatOps that can be used in the Engage step where teams can use ChatOps to collaborate over incidents, problems, and changes in real time. This data is also stored for creating knowledge management.

Knowledge management is a key area in Engage, since ITSM systems are the primary repository of most of the information in IT service management. AIOps techniques such as natural language processing and text extraction come in handy to find the relevant information while resolving incidents and executing changes and service requests. The AIOps system can use information retrieval and search techniques to find the relevant information quickly and easily so that the operations teams can resolve incidents faster.

## Feedback

The feedback in the Engage phase is generated through various mechanisms and in various processes. In the incident management process, the closure of an incident triggers feedback that is filled in by the impacted user; similarly, reopened incidents are a feedback mechanism for analytics. Feedback on failed changes or changes that had to be aborted and changes that were executed fully but caused incidents is an important input.

Service requests raised by users also trigger a feedback post-completion and form an input to analytics to understand how well the process is performing.

All this data is fed into the system to be visualized in the visualization layer and analyzed using analytics techniques.

Rather than acting as feedback to an algorithm, the feedback here is used mostly in data analytics for decision-making to improve the overall process.

The Engage ITSM systems orchestrate the entire process, and every step of the process is logged and updated in the Engage phase; however, the actual action performed is under the Act phase, which we will be discussing next.

# Act

The Act phase is the actual technical execution of the task that includes execution of incident resolution, service request fulfilment, change execution, etc. Figure 2-6 offers a visualization of this phase.

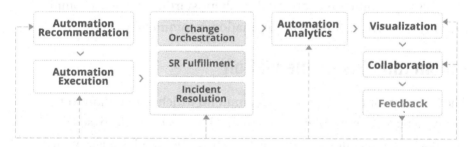

***Figure 2-6.*** *AIOps-driven IT automation*

Thus, all technical tasks executed by the operations team come under this phase.

The completion of the AIOps journey happens with the Act layer; it is here that the incident is resolved, and the system is brought back to its normal condition. AIOps has benefits without this layer as well, where most of the diagnostic and analytics activities are covered under the Observe and Engage sections; however, extending AIOps to Act increases the benefits manifold as organizations are able to not just find problems quickly but are able to resolve them automatically without human intervention.

For the Act layer to work, it is essential that we have the Observe layer implemented and fine-tuned. Without the ability of the AIOps engine to detect anomalies and probable cause and trigger an action, it is not possible for the Act layer to resolve a problem.

Thus, the Act layer is integrated with the Engage and Observe layers to get its data feed and then acts on that data feed to take resolution or other actions on the technical environment. The Observe layer uses AIOps techniques as described earlier to find the probable cause and then create an incident in the ITSM system or the Engage layer; the automation of Act layer can pick up these incidents from the Engage layer and then resolve the incident automatically. To resolve incidents automatically, it needs to know how to resolve an incident. It also needs to understand the incident and the infrastructure on which this incident has happened. We will look at various techniques that are used in AIOps starting with least complex but very effective technique of automation recommendation.

## Automation Recommendation

The first step for resolution is to recommend which automation will resolve a particular problem. This can be done using a rule-based approach or a machine learning approach. In the rule-based approach, each type of probable cause is mapped with an automation, which is fired to resolve the probable cause. In machine learning AIOps approaches, this relationship is not fixed and is probabilistic.

Various techniques like natural language processing and text extraction are used to find the right automation for resolving a problem and then recommending that as a solution for the probable cause identified in the Observe layer.

The automations are generally static in the AIOps domain; however, there are newer technologies that use advanced machine learning techniques to club together runbooks that can be chained to resolve problems, thus creating new automations on the fly using machine learning. Tools like DryICE iAutomate provide these advanced features along with out-of-the-box runbooks and pretrained models to significantly enhance the automation recommendation capabilities.

Automation recommendations provide a confidence score for the automation. Low-risk tasks can be automatically mapped for execution of the recommendation, and any high-risk execution tasks can have a human in the middle approach where a human operator validates the recommendation before it is sent for execution.

## Automation Execution

Automation execution is the actual act of executing on the recommendation generated in the previous step. Thus, once a probable cause has been mapped with an automation, the automation is triggered to resolve the problem.

The execution layer can be built in the AIOps platform or can leverage existing automation tools available in the environment. The automation execution engine provides feedback to the AIOps tool on successful or unsuccessful execution results.

The automation can be triggered using a variety of tools including runbook automation tools, configuration management tools, provisioning tools, infrastructure as code tools, robotic process automation, and DevOps tools. Most organizations have multiple automation tools, and the relevant ones can get integrated as the automation execution arm to be fed by the automation recommendation engine.

Automation tasks can be simple ones like running a PowerShell or shell script to reboot or restart services or can encompass more complex tasks that involve complex workflows and even involves spinning up new instances and infrastructure.

Most AIOps platforms today operate only at the Observe layer and do not have automation execution or recommendation as part of the toolset; however, there are tools like DryICE iAutomate that provide automation recommendation and execution capabilities bundled with out-of-the-box workflows so that organizations can leapfrog and reach a higher level of maturity quickly.

Outside of the AIOps platforms, there are countless scenarios where it is possible to use AIOps disciplines as a trigger to an existing automation. Most organizations have Python or PowerShell scripts that can automate routine remediation workflows, such as rebooting a virtual machine. Expect these types of predefined automations to be a substantial portion of your intelligent automation portfolio, and reuse automation assets with AIOps solutions to increase the value of both the AIOps analysis and the automation development.

Automated execution can span different types of use cases including incident resolution, service request fulfilment, and change orchestration. Some of them are more tuned for a probabilistic scenario, while others can use rule-based process workflows.

One should take care while triggering low-confidence recommendations, or the failure of automation may lead to further problems. In such situations, it may be prudent to use other techniques such as diagnostics runbooks. The low-confidence automations can also mandate a doer checker process whereby the actions of the system are evaluated by a human operator before being fired in the system. A combination of a rule-based system with confidence-based recommendations would best work in certain situations, and thus an intelligent call needs to be taken based on the environment and the risk associated.

## Incident Resolution

Incident resolution is one of the types of automated execution. This is one area that tightly integrates with the Observe phase. Thus, the output of the observe phase, which is a probable cause, becomes an input to the automation assessment, and if there is an automation available for that particular root cause, it can be triggered automatically or using a human in the middle approach to resolve the problem.

Incident resolution is an area where machine learning techniques of recommending an automation can be used effectively and will provide better results than rule-based systems.

Incident resolution is the primary area where probabilistic machine learning technologies in AIOps play a role.

## SR Fulfilment

Service request fulfilment is an area where users request particular services, which are logged into the ITSM system as a service request. The service request is then fulfilled as a series of tasks that get executed automatically or by human agents, or by a combination of both automation and human actions.

Since service request tasks are mostly definitive in nature and there is no ambiguity on how this task needs to be executed, the role for machine learning technologies is limited.

Service requests are fulfilled through a series of step-by-step processes termed *tasks*. At each stage of execution, the requester is kept updated on the progress of his request, and on completion of the task, the requester is notified of fulfilment and a way to access the fulfilled deliverable.

Service requests can be for software systems, or they may be for hardware that needs to be delivered; for example, the delivery of a laptop to a new employee is a service request, which actually goes through a physical fulfilment process and hence cannot be fully automated. Automation here is about integration of service request systems with procurement and ordering systems so that the request can be automatically forwarded to third-party partners or vendors who would take this up for fulfillment by way of shipping a laptop/hardware.

Software deployment tasks are fully automated using software delivery platforms for service requests, which include things like deployment of end user applications like Microsoft Office on laptops.

The service requests are initiated by the requester from a service catalog. The service catalog is similar to the shopping cart where you can order goods and services from online marketplaces like Amazon.

The role of machine learning in this area is around cognitive virtual assistants, which can provide an intuitive chat or voice interface to users rather than a web portal or catalog. This makes it easier for users to converse in natural language, arrive at the right catalog item, and then order it all from a chat interface. The cognitive virtual assistants are integrated in the Engage layer and raise a request after the confirmation from the user. The virtual agents can also be used by the requester to track the progress of fulfilling the request.

Cognitive virtual assistants internally use natural language processing and understanding along with various machine learning and deep learning technologies to decipher the intent of the user and provide appropriate responses.

The primary use case for cognitive virtual assistants is in the area of service request; however, similar functionality and use cases are equally appliable in incident management and change management.

## Change Orchestration

Similar to service request fulfilment, the changes are granularly planned and comprise a series of tasks to be performed by different teams.

There are additional tasks around change review where a change advisory board comprising technical and business stakeholders reviews all aspects of the change and approves it.

There are other steps such as reviewing the change test plans, rollback plans, etc., and at each stage different stakeholders may be involved in the review and analysis of the change.

Once everything is reviewed and the change is ready to be executed as per schedule, the change is executed in a step-by-step fashion by various teams involved in the technical execution.

Thus, change orchestration is a well-scripted process that involves well-defined steps and tasks at each stage, and rule-based systems have been used to run this process for ages. Since change orchestration tasks are mostly definitive in nature and there is no ambiguity on how this task needs to be executed, the role for machine learning technologies is limited.

There are a few areas where machine learning or analytics technologies can be used in change orchestration.

Change scheduling and conflicts is one such area. When a change is scheduled, it involves infrastructure, platform, and application components that form part of the change; the change is also scheduled on a particular time on a particular date. Analytics techniques can be used to find out if there are items that are getting impacted because of the change and if there is conflicting or overlapping changes that impact connected devices or systems. This data can be analyzed by overlaying the topology and configuration data with the change schedules, and this information can be used by the change advisory board to better analyze the change and its impact and may result in the rescheduling of changes in the case of conflicts.

The other area is around change risk analysis; we are aware that every change carries a risk to the application and infrastructure and can result in downtime, and the predictive analytics techniques can be used to find out the risk of the change based on components involved, the complexity of the change, and the risk analysis data from previous such changes. This predictive analytics component from AIOps can come in handy for such analytics and providing this additional information to the change advisory board and technical execution teams.

Cognitive virtual assistants internally use natural language processing and understanding and can also be used in the change management process for technical and process teams to collaborate and find information about the change using the intuitive NLP capabilities of the cognitive virtual agents.

We have covered some of these elements in the change analytics section in Engage as well.

## Automation Analytics

Automation analytics is an important area in the execution space. Though most of the analytics is relevant in the Engage phase, automation execution generates its own valuable data that needs analytics.

Some of the data generated by the automation is used to further improve the accuracy and efficiency of the automation system. The other data is used for reporting and analytics of how the current state of automation is performing in the organization.

Generally, the following are important automation KPIs:

- Automation coverage

- Automation success rate

- Automation failure rate

- High used use cases

- Low used use cases

- Failure cause analysis

- Low automation areas

## Visualization

Just like the visualization and dashboarding that we covered in the Engage phase, the automation phase also logs its data into the system. We mentioned a few important parameters that are important for automation teams to track, and these need to be visualized using dashboarding technologies to provide a bird's-eye view of how automation is performing and run some level of analytics to run continual service improvement on this data.

## Collaboration

Collaboration in automation is largely delivered through the Engage phase since all the activities are logged in the IT service management systems. Thus, collaboration between various teams and people involved in incident, service request, and change orchestration is done in the Engage phase using ITSM tools.

However, with AIOps ChatOps, cognitive virtual agents become the core of collaboration for various teams where they can interact with the right stakeholders as well as get the required data from the ITSM systems, which is required during the Act phase. Thus, real-time collaboration during the actual activities in the Act phase is done using ChatOps where humans interface with other teams and with machines to analyze the data and take appropriate decisions.

## Feedback

This information is available in the AIOps system to analyze the efficiency of the automation execution. Feedback from automated resolutions is an essential input into the AIOps system. Success and failure of automation scripts is an important learning data for the machine learning algorithms. This data helps the system to improve its accuracy scores and to change the confidence scores of the various automation engines and scripts.

Operators confirm the actions that the AIOps algorithms are suggesting, and thus the human input provides training to the AIOps algorithms where they can improve on their model and change the confidence scores based on human feedback.

With time the AIOps engine becomes better tuned to the environment that they are operating in by learning from the actions of human operators. The confidence scores of various resolutions become high enough to move them to fully autonomous mode where the human in the loop is no longer needed for certain use cases. This fulfills the promise of AIOps to turn operations into NoOps, by leveraging the autohealing capabilities of AIOps technologies. Expansion of fully automated use case library and its success rate gets greatly impacted by the knowledge of the application blueprint and its relevance or its business impact level (BIL), and that's where application discovery becomes important, which we will be discussing next.

# Application Discovery and Insights

To manage business transactions' key performance indicators (KPIs) and to guarantee business process service level agreements (SLAs), enterprises also need powerful full-stack analytics.

These analytics need to automatically map business transactions (such as orders, invoices, and so on) to their application services (web server, application server, databases, and so on) and to the supporting infrastructure (compute, network, and storage), as shown in Figure 2-7.

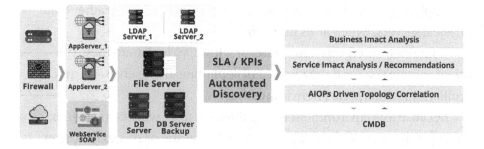

***Figure 2-7.*** *AIOps for business and service impact analysis*

This must be done in real time across distributed, hybrid IT environments. Without it, they will be forced to engage in extensive and complex troubleshooting exercises to triangulate hundreds of thousands, if not millions, of data points. The time required to do so can negatively impact the uptime and performance of key processes such as e-commerce, order to cash, and others.

# Making Connections: The Value of Data Correlation

The app economy is upon us, and businesses of all stripes are moving to address it. In this age of digital transformation, businesses rely on applications to serve customers and improve operations. Businesses need to rapidly introduce new applications and adopt new technologies to become more agile, efficient, and responsive.

As part of these efforts, businesses are employing cloud-based solutions, software-centric and microservices architectures, and virtualization and containers. But these new architectures and technologies are creating challenges of their own.

Some business applications today are hosted on public clouds, and enterprises tend to have no, or very limited, visibility into those clouds.

Applications increasingly deployed on virtual machines rather than physical servers, which adds more complexity.

Containers often exist alongside, or within, virtual machines. The use of containers—and the number of containers themselves—is quickly proliferating enterprise IT environments.

Because this environment is very different from what came before, the application performance tools created a decade or so ago no longer apply. Tools that consider only the application—and not the underlying infrastructure—fall short. These tools must collect and correlate information about the application itself and about the underlying infrastructure. This should include data about application server performance, events, logs, transactions, and more. The compute, network, and storage resources involved in application delivery also need to be figured into the equation.

# Summary

In this chapter we covered the different layers of AIOps, i.e., Observe, Engage, and Act. The three core foundation pillars of AIOps along with the core analytics techniques of event correlation, predictive analytics, anomaly detection, and root-cause analysis were covered in depth. Each function within the Observe, Engage, and Act layers was covered in depth by providing real-life practical examples so that AIOps and its functions can be demystified and correlated with on-the-ground activities being done by operations teams today. In the next chapter, we will cover challenges that organizations face while deploying AIOps.

# CHAPTER 3

# AIOps Challenges

Though IT teams are optimistic and moving full steam ahead on implementing AIOps technologies, there are some challenges that hinder the value realization and implementation of these technologies. This chapter explains such challenges in more detail so that organizations can plan for them while implementing AIOps. Figure 3-1 lists the key challenges in the AIOps journey that we will be exploring further.

© Navin Sabharwal and Gaurav Bhardwaj 2022
N. Sabharwal and G. Bhardwaj, *Hands-on AIOps*,
https://doi.org/10.1007/978-1-4842-8267-0_3

*Figure 3-1.*  *AIOps challenges*

# Organizational Change Management

AIOps is a transformational theme that cuts across various processes such as ITSM, monitoring, and runbook automation. It also cuts across various teams including the command center, the service desk, resolution teams, automation, SRE, and DevOps. To successfully deploy and realize value, you need effective change management at the organizational level, along with support and sponsorship from the management to see this cross-functional change through. Organizational hierarchies and functional divisions get in the way of successful and effective deployment of AIOps; thus, it is important to engage organizational change managers to run the project through various stakeholders and get management buy-in for the project.

Typically, the previous teams are structured in different functional hierarchies, and they collaborate on an as-needed basis on initiatives through defined processes and policies. AIOps is a disruptive change that impacts all of these teams drastically.

To be able to get the project off the ground and be successful, it is important that this is driven as a program of organization-wide significance and monitored and governed at the highest levels.

Team structures, processes, policies, and communication mediums need to change to adopt AIOps in totality and realize the maximum benefits by deploying these technologies across the entire value stream.

Thinking of AIOps as just a technology change that can be dropped on top of existing technologies and processes without re-engineering is an expectation that needs to be tapered down, and organizations should realize that they are moving to a different system and method of doing IT operations that requires changes to the current structures, teams, and processes. Therefore, this needs to be handled in a manner that is process oriented, systematic, and nondisruptive to existing operations and provides clear guidance to the people who are going to be impacted by this organization-wide change.

# Monitoring Coverage and Data Availability

There are organizations where monitoring and observability are in place, and they can proceed smoothly to deploy AIOps technologies and reap the benefits that AIOps provides. However, there are organizations where the basic monitoring coverage is not complete, and many infrastructure and application components are not monitored effectively. Thus, due to lack of data, deploying AIOps would not be able to provide probable cause or root cause for those application or infrastructure elements. After all, machine learning systems are based on data availability and accuracy of data. There are other challenges where the coverage may be complete, but the monitoring parameters are incorrectly configured and do not provide complete and accurate data for effective functioning of the AIOps algorithms.

Without monitoring and observability data in place, expecting AIOps to solve basic monitoring or observability problems would result in a failure of the initiative and AIOps getting the blame for no fault on the AIOps side either from a technology or process perspective.

IT operations needs to be able to measure their maturity and coverage of systems and processes from monitoring, observability, ITSM, and automation standpoints before embarking on an AIOps journey. This can be done either in house or by engaging consultants who are well versed with this domain and can provide an outside perspective to where the organization is and compare the systems and processes with companies that are leaders in this space. This exercise will result in a maturity journey where there may be projects and programs that need to be run to prepare the organization for AIOps. For example, if the monitoring is not comprehensive, a separate program to set this right and provide comprehensive coverage may get initiated as a preparatory step to bringing in AIOps.

# Rigid Processes

There are organizations where the event management process and procedures are rigid and it is difficult to change them. The processes have been created based on the existing technologies at that time, which means they are not appropriate for probabilistic models and are hardwired for each type of event. Deploying AIOps needs process changes to accommodate for probabilistic decision-making rather than rule-based decision-making. At times organizations end up deploying AIOps tools but deploy rule-based logic rather than machine learning, thus reducing them to non-AI-based traditional event correlation systems.

There are organizations that want to adopt AIOps but do not want to change the current processes that are set up based on technology that evolved a decade back and largely remained consistent. AIOps needs a mindset change where you may not be able to test the system granularly with all the data and define test cases that replicate all real-life scenarios or are 100 percent predictable. The AIOps system is not a stagnant rule-based system that will give the same result if the same data input is provided. The system is ever learning, and thus the models keep changing based on the data that the AIOps system is ingesting and the learning that humans are providing to the system.

Expecting AIOps to work in the same fashion as traditional systems creates incorrect expectations from an AI system. AI-based systems work on data and probability, and thus the results may vary from organization to organization, from infrastructure to infrastructure, and from time to time based on the unique data that is generated in each organization.

Fine-tuning of the system ensures that the confidence scores and accuracy of the AIOps system is maintained and improved on an ongoing basis.

# Lack of Understanding of Machine Learning and AIOps

A few organizations may not have expertise and experience in machine learning and AIOps, and they may struggle to understand and accept the benefits that this technology transformation can provide. In such situations, it is important to expose the stakeholders to the new technology through webinars, trainings, and proof of concepts to make them aware of the new methods and techniques.

A lack of understanding of how AIOps works is one of the most important aspects to consider while moving ahead with an AIOps project.

Typically, the monitoring and management teams are experts in their own domain and have worked on technologies that have serviced their needs for decades. The teams are generally comprised of subject-matter experts in the monitoring and observability domain as well as experts in various technology domains such as networks, cloud computing, storage, security, and data center; however, the teams may not have any expertise in AI and machine learning since this was not needed for them to run IT operations.

Thus, there is a gap in capability and expertise that is needed to make the best use of AIOps systems. One does not need deep-level expertise in AI or machine learning to be able to use the AIOps system; however, they do need exposure and beginner-level skills in AI and machine learning to be able to understand what the AIOps system is trying to do and how to utilize it to its potential.

We sincerely hope that this publication will bridge the gap and expose the monitoring and management subject-matter experts to enough AI and machine learning that is needed for them to successfully implement and run AIOps-based systems.

# Expectations Mismatch

There are times when there is an expectation mismatch and the AIOps tools vendor makes promises on value realization without understanding the bottlenecks of processes, systems, functions, and technology limitations in an enterprise. The tools may provide extensive capabilities, but without changing the processes and the functions, the realization of benefit may not happen. A typical case in point is collaborative ChatOps between the Dev and Ops teams. If an organization is siloed and does not have DevOps processes in place, deploying an AIOps feature for ChatOps and collaboration is not going to help bridge the gap. This needs to be handled elsewhere in the project, and the organization needs to embrace DevOps before the tool's capabilities can be used effectively by the team.

Expectations mismatch stems from some of the areas that we covered in earlier points. For example, a lack of understanding of AIOps and AI and machine learning in general will result in an expectations mismatch as the teams might be expecting AIOps to be a magic wand that will solve all their problems. Alternatively, the operations teams may want the AIOps systems to just replicate the features and functionality of their rule-based systems, which will again result in a mismatch in the capabilities of the tools versus the expectations.

Another expectation that security and compliance teams may have from the new AIOps system getting deployed is to have 100 percent accuracy on the insights that the system is providing, and that is an incorrect expectation to have from a machine learning system that works on probabilities and not rules.

It is important to have all stakeholders as part of the program so that the expectations from different stakeholders can be tuned and managed based on what AIOps is and what it delivers for IT operations in order to avoid understanding gaps of the AIOps domain, technology, or processes.

# Fragmented Functions and the CMDB

AIOps works best when the entire estate right from the business process to the application, infrastructure, and network provides the data to the AIOps platform. There are challenges if the monitoring systems and the teams for different technology domains are fragmented and they are unwilling to be onboarded to new integrated process and function. AIOps can easily handle the fragmentation in monitoring systems; however, different configuration management systems and a lack of configuration management database pose problems when correlating events.

The AIOps systems will work best if the machine learning algorithms are provided with more and correct data. Getting accurate topology information vastly improves the accuracy of the AIOps systems to pinpoint problems and provide a root cause. Without this data, the system is running blind and trying to figure out the event correlation based on timestamp data as well as statistical techniques to correlate various events, which may not be as accurate.

It is not necessary that a complete CMDB is provided to the system; however, getting topology information into the AIOps system provides for much better event correlation and root-cause analysis.

If it is difficult to get topology or configuration information for the entire landscape, one can start with critical applications and infrastructure and slowly expand the coverage to the other areas.

# Challenges in Machine Learning

AIOps uses a combination of rule-based, machine learning–based, and topology-based systems, and the data from different monitoring and management systems as well as their configuration is fairly unique in each environment. This poses a challenge for the machine learning models as the models need to understand different events and provide automated analytics alerts. Fine-tuning the model based on each unique environment may become a tedious process.

Machine learning works best when accurate and complete data is provided, and that may become a challenge if monitoring systems are not correctly configured or events and logs are providing data that is incomplete or incorrect.

Other areas of concern in machine learning are around explainability where it is difficult to pinpoint why a decision was made by the machine learning system. Newer AIOps systems have explainability aspects built in so that one can do a deep dive and find out why the AIOps system recommended a particular root cause or flagged a particular anomaly.

Machine learning systems are also prone to data drift, and changes to the monitoring systems or landscape will have an impact on the AIOps system; therefore, having topology information helps to overcome these aspects.

# Data Drift

All machine learning models have the challenge of data drift. This becomes more pronounced if the organization adopting AIOps is simultaneously working on multiple transformational projects like cloud migration. Since the monitoring tools and data from systems is changing drastically, the data will drift, and the model's accuracy based on historical data that is available will reduce. The data drift in AIOps needs to be carefully handled through processes and analysis to ensure that the models are accurate with current data.

As the data drifts drastically because of transformation projects, the events, symptoms, and metrics will change. This will mean that the data prior to the transformation is invalid and cannot be clubbed with the new data to generate anomalies, dynamic baselines, or prediction on capacity.

One has to be careful to make changes to the data sources or change the data itself as this will make the current deployed models obsolete.

Experts running the AIOps systems should have a basic idea about the data and how it impacts machine learning models. We hope that by going through this hands-on guide and implementing some of the hands-on exercises one would be able to better grasp the relationship between changes in data and the generated models and use this knowledge to ensure that the AIOps systems are utilized effectively.

# Predictive Analytics Challenges

Predictive analytics is a challenging field for any machine learning or deep learning system; it can never be 100 percent accurate. Knowing exactly when a system will fail is unrealistic since failures have an element of chance and thus are impossible to predict with the expected accuracy.

As mentioned earlier, AIOps systems use predictive analytics, and the systems work on probabilities and may generate false positives where incorrect events are flagged as probable root cause or false negatives and real impacting events are flagged as noncritical.

This is similar to the way that these algorithms work in other areas, for example, when a genuine credit card transaction is declined and flagged as fraudulent or fraudulent transactions pass through the analytics engine and are approved.

Other challenges in predictive analytics are around the seasonality of the data that modern machine learning technologies cover; however, long-term seasonality data may not be available, and thus the system will not be able to learn seasonality until it has gathered enough data to do so. For example, to be able to generate seasonality-based predictions for annual seasonality, you would need data for multiple years.

# Cost Savings Expectations

Customers often have expectations around cost savings; however, just by deploying event management AIOps there will be limited cost savings. There may be significant impact on the user experience and customer satisfaction, but without automation, the cost impact may not be as per the customer's expectations. Thus, tools like iAutomate that provide automation capabilities are deployed alongside the event correlation platforms to gain significant cost savings by automating the entire value stream from alert detection to remediation of the problem.

By leveraging AIOps, organizations will reach a higher level of maturity in operations, there will be less noise in the system, and resources will be able to handle a higher volume and improve on the SLAs that are committed to business.

One will see a lower mean time to respond and mean time to resolution by deploying these technologies.

There will be higher availability of systems and a higher level of customer satisfaction from both business users and end users.

All these are measurable and tangible benefits that AIOps brings to the table. However, to be able to reduce cost, there are other factors at play. As an example, the organization may be understaffed or has already applied all cost levers and has a bare minimum team staffed to work as per the service levels committed. In these scenarios, there may not be cost reduction.

The AIOps systems also need to be paid for and maintained on an ongoing basis. This will result in additional costs that an organization will incur to move up the maturity level and provide better services to their customers. Any resource savings that may accrue because of leveraging these systems need to be first adjusted against the investment and operations expenses that are needed to set these systems up and manage them on an ongoing basis.

Thus, organizations should create a business plan and look at all costs and savings before setting expectations on costs and savings.

# Lack of Domain Inputs

AIOps needs domain inputs from technology experts who are operating a particular environment. This can be in terms of providing feedback on the output of the AIOps platform or documentation and knowledge repository. A lack of buy-in from the technical experts results in no supervised training for the AIOps engine, which then means that the AIOps engine falls back on unsupervised learning to provide results. Without the feedback loop from domain experts, the continuous learning of the machine learning models does not happen, and subsequently the accuracy of the system does not reach its potential.

There are various challenges in getting domain experts on board on this transformation journey. It could be a plain simple lack of understanding of how AIOps systems operate or a preference to work in the current mode of operations and continue with the processes that people are used to from decades ago.

Organizational change management and AIOps process setup phases should have a buy-in from all domain experts so that everyone contributes to the project and is able to visualize common successes.

Identifying stakeholders early on, setting up KPIs for collaboration, and incentivizing the team for the success of the project are a few things that organizations can do to have a successful AIOps project.

# Summary

In this chapter, we covered the challenges that organizations face while deploying AIOps and how they need to be mitigated. It is essential for any organization implementing AIOps to understand the challenges beforehand and plan for their mitigation to successfully implement an AIOps project. In the next chapter, you will see how AIOps enables SREs and DevOps teams to work efficiently and how AIOps aligns to the SRE and DevOps principles.

# CHAPTER 4

# AIOps Supporting SRE and DevOps

Just like AIOps is disrupting the technology and processes for IT operations, there are other transformational changes that are happening in the IT operations space.

DevOps as a movement has gained momentum and now cuts across the development, infrastructure, and operations worlds. With cloud infrastructure and cloud-native applications becoming the norm, DevOps is expanding its coverage to include end-to-end workflows right from development to deployment. AIOps supports the DevOps model and promotes collaboration across the development and operations teams.

Similarly, another movement that is closely related but has a slightly different take on operations is site reliability engineering (SRE), where the focus is on managing and operating applications and platforms with a focus on reliability, availability, and automation.

This chapter explains how AIOps, DevOps, and the SRE model are complementary and how one can underpin the DevOps and SRE services by using AIOps technologies.

Let's begin with the overview of the SRE model and DevOps.

© Navin Sabharwal and Gaurav Bhardwaj 2022
N. Sabharwal and G. Bhardwaj, *Hands-on AIOps*,
https://doi.org/10.1007/978-1-4842-8267-0_4

# Overview of SRE and DevOps

AIOps is not the only discipline that is changing the way IT is run. DevOps, Agile, and SRE are the other disciplines that are transforming IT operations. Agile and DevOps have driven a cultural shift in organizations where agility and speed of delivery coupled with collaboration between development and operations resulted in benefits to enterprises that adopted them. Similarly, AIOps is changing the way collaboration is done between the development and operations teams using data and analytics and providing insights and knowledge that weren't available earlier. With ChatOps and knowledge management, AIOps is providing the technology foundation on which the development and operations team collaborate. Both AIOps and DevOps together are taking organizations to even higher levels of automation and maturity.

AIOps is providing access to technologies that can bring data, data analytics, and machine intelligence together to make informed decisions and perform automated actions by collecting and analyzing data. Figure 4-1 shows the DevOps process from Plan to Deploy and the various AIOps functions that we discussed in earlier chapters.

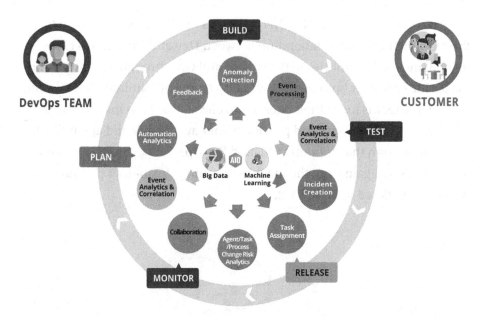

**Figure 4-1.**  *AIOps and DevOps, better together*

AIOps and DevOps should be adopted across the enterprise to reap their benefits; deploying them in departmental silos wouldn't make organizations move up the maturity ladder. Though DevOps implementations generally follow a step-by-step approach to implementation, AIOps needs all the data that is in the monitoring and management space to be able to better correlate data and provide insights and analytics.

AIOps eliminates noise and thus helps in making DevOps more effective. With AIOps in place, false alarms are reduced in the system, and this reduces the wasteful work for DevOps teams to analyze the false positives in the system. DevOps teams get single alerts rather than multiple alerts from various systems in the AIOps model along with probable root cause so that they can focus their energies on resolving the problem rather than trying to decipher and analyze so many alerts that are getting generated because of an issue in an application or infrastructure.

Thus, AIOps supports DevOps and helps the DevOps teams realize their vision of greater collaboration between Dev and Ops. AIOps helps in breaking the boundaries between Dev and Ops through a platform and makes life easier for both the Dev and Ops teams by eliminating waste and helping the teams focus their energies to elevate the DevOps processes to a higher level of maturity and provide higher availability and agility. AIOps is the foundation that supports SRE and DevOps. Thus, the DevOps and SRE models can gain from the machine learning capabilities provided by AIOps, as depicted in Figure 4-2.

***Figure 4-2.*** *AIOps collaborating between SRE and DevOps*

In today's world, where so many essential business tasks have become digitized, IT teams must deal with constant change while ensuring zero downtime.

In the modern digital IT operations world, the DevOps teams work together, each on their own microservice. The DevOps teams are supported by SRE teams or embedded individuals with the DevOps team with the primary SRE role of maintaining the high availability of applications. The site reliability engineers provide insights to the DevOps team to improve the architecture and code of the applications by analyzing the operations feeds.

The challenge for SREs is to improve the stability, reliability, and availability across disparate systems, while application teams are delivering new features at a rapid pace. To achieve their targets, site reliability engineers have to be one step ahead of the outages and resolve incidents quickly. However, the lack of AIOps tools results in teams getting overwhelmed by noise, and it becomes difficult to isolate the root cause and provide immediate analysis and recommendation.

Analyzing alert data manually is more and more becoming an impossible task. Taking huge dumps of alert data and using Excel and other BI tools to analyze data is no longer going to work out since the monitoring and management data is humongous and in various formats. It is important for the SRE teams to be able to remove the noise and focus on the alerts, which are the root cause of incidents.

With multiple and geographically dispersed teams, collaboration also becomes a challenge. How do SRE and DevOps teams with ownership of different microservices that are mashed up together create an application and resolve incidents? Where do they get the data, the visualization, and the collaboration tools to run operations? AIOps comes to the rescue by providing the DevOps and SRE teams with the tools and technologies to run operations efficiently by providing them the visualization, dashboards, topology, and configuration data, along with the alerts that are relevant to the issue at hand. Thus, AIOps provides a unique solution to address operational challenges.

As a result, SRE teams are adopting AIOps tools to help address these challenges, including the adoption of AIOps for incident analysis as well as remediation.

Here are a few questions that an enterprise should ask about their SRE operations to arrive at the need for AIOps:

- Do you have effective collaboration tools for DevOps and SRE teams?

- Does your organization use automation and tools to improve resiliency?

- Are your SRE teams able to manage the SLAs and error budgets?

- Are your SRE teams getting the right alerts, or are they overloaded with false positives?

- Are your SREs able to quickly find root causes using automated mechanisms?

- Are you using ChatOps and generating knowledge while running operations?

- Are your SREs using automation for incident resolution and configuration changes?

Based on the environment complexity, processes maturity, and investments made on tools and solutions, different organizations have defined and implemented SRE principles that may vary. In the next section, we will be discussing best practices and SRE principles in relation to AIOps that can be widely adopted by organizations and how AIOps supports the key principles of SRE model.

# SRE Principles and AIOps

Site reliability engineering has gained traction as a domain and skill in recent years. With application and infrastructure complexity increasing because of many architectural choices, availability and resilience are essential in both architecture and operations. The SRE model is built on the following principles; you will see how AIOps enables most of them.

# Principle 1: Embracing Risk

Embracing risk means weighing the costs of improving reliability and the impact it has on customer satisfaction. No service can be reliable 100 percent of the time. There is also a cost trade-off with reliability; after a certain point, adding more reliability would mean doubling the cost or even increasing the cost multifold. As an example, to support 99.95 availability versus supporting 99.999, the cost difference can be multiple times. Thus, there has to be a balance between the goal for reliability and the cost associated with it.

SRE service level agreements around availability and response time coupled with error budgets support this principle where they are free to manage availability within the SLA ranges. The SREs have a right to reject changes to applications or infrastructure if they are running low on error budget. Also, the SRE focus is on getting the services up and running as quickly as possible, which may involve taking some level of risk with quick decision-making.

AIOps helps SREs in this principle by providing them with all the data to measure SLIs and SLOs and then aggregate them under SLAs. The AIOps tools enable quick resolution of incidents using automated mechanisms that can be fired by the SREs to resolve availability problems. The AIOps models and analytics are not deterministic but probabilistic and thus carry an element of uncertainty and will never be 100 percent accurate and thus align with the SRE model.

# Principle 2: Service Level Objectives

The observability data forms the basis for service level indicators that provide data on things like availability, response time, etc. These metrics are then aggregated under service level objectives (SLOs). Service level objectives are set to the point where the customers will feel dissatisfaction with a service. The service level indicators and objectives will be different for different types of business requirements and users. As an example, we have a 100 percent availability expectation when it comes to mission-critical applications or internet-scale applications like search and email. However, we may not have the same expectations from less critical systems like a timesheet application. Similarly, our requirements for response time from a mail or search engine application versus an ERP application are different. Service level objectives take the business and customer context and apply it on top of the service level indicators to arrive at the objective or target that the SRE team is willing to live with. An example could be 99.95 availability.

Service level objectives are for a timeframe. For example, an objective could be to meet the 99.95 availability goal on a monthly basis.

SLOs also leave room for an error budget. Whenever a failure or degradation affects the service, the error budget decreases. Thus, in the previous example, .05 percent is the error budget available.

It is difficult to achieve granular SLIs and SLOs and measure them accurately without AIOps. With basic monitoring and fragmented monitoring tools working in silos, it is extremely difficult to calculate and arrive at application availability or business process availability. Thus, AIOps is essential to provide the correct data on the availability and response time of an application or business process to enable the SREs to have granular and correct data to arrive at the right SLIs and SLOs.

# Principle 3: Eliminating Toil

Eliminating toil means reducing the amount of repetitive work a team must do.

This is an extremely important principle for SREs; it differentiates the SRE model from other operations models that do not have metrics to measure and eliminate toil. SREs carry targets for toil elimination, and for this there are two important elements. One is the ability to measure toil in the system, and the second is to eliminate toil rapidly so that the teams can focus on more value-adding work.

SREs eliminate toil by automating routine and standard work that does not need human intelligence for every transaction. Things like health checks, routine checks, reporting, automated monitoring, and autoremediation for known errors are some of the tasks that SREs work to automate.

Another way SREs eliminate toil is by creating guides or standard operating procedures so that the knowledge base is enriched and can be quickly searched and used whenever required.

AIOps comes to the rescue here in more ways than one. Though not many AIOps products support algorithmic or machine learning–based automation systems, but there are a few upcoming products like iAutomate that use machine learning technologies to manage automation work. AIOps products provide capabilities where the SRE teams can use algorithms to estimate the amount of toil and keep track of automation and its benefits using analytics to identify, automate, and report on toil. The tools also come with built-in automations that can be used easily by the SREs to eliminate toil rapidly without having to create automations from scratch.

# Principle 4: Monitoring

Monitoring or observability is the key to getting data from systems and applications. Without monitoring and observability, the availability, performance, and response time of applications and business processes cannot be measured. Monitoring looks at events, metrics, logs, and traces to provide meaningful data that can be analyzed and used for reactive and proactive actions to provide higher availability.

Monitoring and observability tools provide the raw data for the AIOps tools that can then use this input and apply algorithms and machine learning techniques to provide deeper actionable insights on this data.

The most common metrics focused on for reliability are these four golden signals:

- *Latency*: The time it takes for a service to respond to a request

- *Traffic*: The amount of load a service is experiencing

- *Error rate*: How often requests to the service fail

- *Saturation*: How much longer the service's resources will last

All these golden signals are analyzed by the AIOps tools to provide forecasting to see if something will break in the near future. This removes noise and selects actionable alerts and supporting data so that the teams can focus on the right alerts without wasting time to find the root cause. Advanced statistical techniques are used to create causality maps that show which component failure resulted in failure of the system. Pattern matching techniques are used to find relevant information in log files, which is then provided to the site reliability engineers to do a deeper analysis.

AIOps tools greatly enhance the SRE function by providing them with the right information at the right time without having to dig through multiple records to arrive at a conclusion on the root cause. The dashboards created in AIOps provide topology, alerts, and knowledge articles so that the SREs can collaborate with development teams.

Using AIOps, the SRE and DevOps teams are able to visualize the entire architecture of an application using the topology and discovery data. Then this data is overlayed with the events that are getting generated from individual layers and components. This gives a bird's-eye view to the entire application landscape and makes it easier for SREs to use this data and visualization in both operations as well as architectural transformation decisions.

The AIOps tools also help the SRE teams in understanding the application from an infrastructure-up perspective by getting the data from infrastructure monitoring tools and overlaying that with real user-monitoring data from application monitoring tools. This helps the SRE teams to narrow down on the root cause, which can be at the user end-level or in the network, infrastructure, or software code. The visibility provided by the AIOps tools helps SREs arrive at the root cause much faster.

Root-cause analysis done using machine learning technologies aids in faster response and resolution time and helps the SRE teams in maintaining a high level of availability and manage their SLAs. This results in increased productivity of the operations teams and higher customer satisfaction scores.

# Principle 5: Automation

Automation means creating ways to complete repetitive tasks without human intervention. This helps free up teams for higher-value work.

This principle is an extension of the eliminate toil principle where everything that creates repetitive work for the team is eliminated. Automation helps to achieve the elimination of toil.

The SREs work on many automation levers.

- *Incident response*: The SREs respond to incidents that have or are likely to impact SLOs. This is an important function for the SREs. The SREs automate incident response by using scripts and creating automation for simple scenarios. They can also leverage tools to deploy out-of-the-box automations for incident response.

- *Deployment*: The SRE teams automate the deployment of monitoring and other applications as well, depending on how the function is structured in an organization.

- *Testing*: This is an important element of SRE work, where automated testing for infrastructure and applications is used by SREs to find resilience issues. SREs use various methods and tools.

- *Communication*: This is an essential ingredient in any operation. SREs use ChatOps tools to collaborate and communicate about issues and development work along with incidents and other actions, and tools like Slack are used by SREs for real-time communication.

AIOps tools help in the automation of monitoring, root-cause or probable-cause analysis, and remediation of incidents. The AIOps tools also provide embedded or integrated ChatOps to facilitate communication between SRE members and development teams on a real-time basis where both can have access to a common shared dashboard and alert data to better analyze the situation and take appropriate remedial action.

# Principle 6: Release Engineering

Release engineering means building and deploying software in a consistent, stable, repeatable way. The previous SRE principles are applied to releasing software.

Some of the activities done as part of release engineering are as follows:

- *Configuration management*: Define the baseline configuration and ensure that the releases change the configuration based on defined processes and that the configuration changes are tracked. It also involves defining "desired state configuration" for systems so that there is no deviation from the approved configuration.

- *Testing*: Implement continuous testing and automated testing to ensure the release meets the requirements and definition of done.

- *CI/CD and rapid deployment*: Where possible, automate the release process by leveraging the DevOps principles of continuous integration, continuous delivery, and continuous deployment. This enables consistent, repeatable releases in an automated fashion and increases the agility by providing teams with a system and process to release very frequently in response to business needs or for bug fixing.

Since the AIOps tools today are focused more on the operations aspects, they are used in the development process to ensure that the release has the requisite operations aspects covered during the development process.

AIOps tools help in identifying configuration drift and unapproved changes to infrastructure or applications by providing analytics over log data that is ingested into the AIOps tools. The tools facilitate continuous testing and deployment by providing insights into the availability and performance of the application and infrastructure in the development lifecycle. The AIOps tools support rapid deployment by providing before and after deployment comparison when SREs and DevOps teams are using continuous deployment and deploying to production frequently. Without AIOps, these processes are manual and prone to errors; AIOps helps the SREs to focus on the core aspects of availability resilience and performance while doing the heavy lifting of providing the right data to them.

## Principle 7: Simplicity

Simplicity means developing the least complex system that still performs as intended. The goals of simplicity and reliability go hand in hand. A simpler system is easier to monitor, repair, and improve.

# AIOps Enabling Visibility in SRE and DevOps

SREs advocate an end-to-end approach to reliability; building models is a good way to gain insights into the internal components. SRE advocates a holistic, end-to-end approach to reliability.

AIOps tools bring simplicity to the architecture by ensuring everything is integrated from observability aspects and integrated into a dashboard. AIOps ties together all the monitoring tools and integrates the automation aspects into the mix. Using AIOps simplifies operations and integrates the Dev and Ops teams through collaboration tooling.

# Culture

DevOps is the culture and mindset forging strong collaborative bonds between the software development and infrastructure operations teams. This culture is built upon the following pillars:

- *Constant collaboration and communication*: AIOps provides the tools for DevOps and SRE teams to collaborate and communicate and provides ChatOps for real-time collaboration.

- *Gradual changes*: AIOps gets well entrenched in the overall culture of DevOps and SRE, and the models evolve over time with the changes in applications, infrastructure, and data.

- *Shared end-to-end responsibility*: AIOps tools enable shared end-to-end responsibility through better coordination across the development and operations lifecycle as well as promoting collaboration and communication.

- *Early problem-solving*: AIOps provides for rapid resolution of incidents using automated root cause and analysis and intelligent run book automation execution.

# Automation of Processes

Automation of processes is the key goal of DevOps and SRE teams. Automating routine work and eliminating toil are achieved using AIOps tools, because the automated analysis of root cause, inputs to proactive problem management, and automated resolution of alerts free up the team's time to take on higher-level activities.

# Measurement of Key Performance Indicators (KPIs)

Measuring various metrics of a system allows for understanding what works well and what can be improved. AIOps tools provide in-depth reporting and analytics and help in creating service level objectives, service level indicators, and SLAs.

# Sharing

AIOps tools enable sharing across the value stream by bringing the data from observability tools and providing insights to both development and operations teams. AIOps helps in knowledge sharing by providing knowledge management capabilities where teams can collaborate, create, and share content using the AIOps engine. AIOps enables the SRE teams to deliver faster and better and is a foundational technology on which SRE processes and teams are built.

# Summary

In this chapter, we covered the SRE model and how AIOps enables the SRE teams to deliver on their promise. We also covered how AIOps enables true collaboration between the development and operations teams. We looked at various features provided by AIOps tools and how they support and enable SRE and DevOps principles. In the next chapter, we will cover the fundamentals of AI, machine learning, and deep learning.

# CHAPTER 5

# Fundamentals of Machine Learning and AI

You learned about AIOps, its architecture, and its components in previous chapters, and you will be venturing deeper into AIOps in the next few chapters. Before you start implementing algorithms for AIOps, though, you need to learn a few basic things about machine learning and artificial intelligence. This chapter will cover the fundamentals of artificial intelligence, machine learning, and deep learning. It will then go on to discuss specific techniques that are used in AIOps.

## What Is Artificial Intelligence and Machine Learning?

Artificial intelligence, machine learning, and deep learning are terms that are often used interchangeably. This chapter will introduce each of these terms and detail the differences and overlaps between these technologies.

Most of us have come across AI in some form or another. Today AI is being used in most of the applications that we use every day. From recommendations on shopping sites to finding the most relevant article through a search engine like Google, everything involves AI. We all use

© Navin Sabharwal and Gaurav Bhardwaj 2022
N. Sabharwal and G. Bhardwaj, *Hands-on AIOps*,
https://doi.org/10.1007/978-1-4842-8267-0_5

speech recognition technologies in the form of voice assistants such as Alexa, Siri, etc. All of these are powered by artificial intelligence. What started as science fiction has become part of our everyday life; artificial intelligence has come a long way in a short span of time.

Let's start with understanding artificial intelligence. The term *artificial intelligence* is used to define intelligence that is created by humans and programmed by humans. The AI systems mimic human intelligence and reasoning to make decisions and solve problems. Just like human intelligence resides in the human brain, AI resides in the form of algorithms, data, and models in a machine.

AI has been used to beat humans in games like Chess and Go. AI is also solving problems and creating new use cases for various applications that were not possible before it became available. AI is now being used in diverse fields including military and surveillance applications, healthcare, and diagnostics. Even the legal and governance field is something that has not been left untouched by artificial intelligence. Thus, AI as a technology is finding applications in almost everything.

Artificial intelligence as a term has gained traction since 1950s. It encompasses every type of technology that has an element of intelligence that is artificially created. Machine learning blossomed in the 1980s, and with more computing power being made available, deep learning gained traction starting in 2010. We will be explaining each of these in this chapter.

AI has exploded over the last decade primarily because of newer hardware technologies like GPUs that have made processing faster and cheaper. More powerful processing systems are being made at lower costs, and this has resulted in making the AI algorithms and techniques feasible. Theoretically, the techniques have existed for a long period; however, the technology infrastructure to execute these models at scale did not exist.

# Why Machine Learning Is Important

AI is broadly classified into general AI and narrow AI. General AI as a concept that is still science fiction. General AI is a machine that possesses all the senses that we possess such as sight, sound, touch, smell, and taste and maybe even beyond human senses to see and experience things that are beyond human capabilities. These general AI machines would act just like humans do and possess powers beyond human capabilities. The term *general* in general AI means that just like humans are very general purpose and can do a variety of tasks and learn a variety of things, the machines should be able to do the same.

In contrast, *narrow AI* consists of technologies that perform specific tasks as well as or better than humans. The current state of the technology and implementations are centered around narrow AI. Think of narrow AI as a specialist that can perform one type of a task quickly and with accuracy but is not capable of learning other tasks that are different in nature. Examples of narrow AI are systems that can understand natural language, systems that can translate language, systems that can recognize objects, etc. These systems are created and fine-tuned to do a particular task.

There is a third term called "AI superintelligence," which is used extensively in science fiction to depict an artificial intelligence that reaches a critical threshold of intelligence with which it can optimize itself in progressively smaller timeframes and reach a level of superintelligence. Thus, superintelligence has the ability to transform and optimize itself rapidly by coding itself recursively and improving with each iteration. As the superintelligence improves itself with each iteration, it becomes faster and more powerful, and thus each subsequent cycle takes less time. It is postulated that if such a system becomes reality, it may be able to recode itself millions of times and consume all the available compute power on the planet to create a superintelligence in hours as each successive step takes less time.

Broadly, AI systems have these three key qualities:

- *Intentionality*: The AI system has the intention of making a decision based on real-time analysis of data or based on historical analysis of data. For example, an algorithm for correlation can classify website hits as normal user hits or a DoS attack.

- *Intelligence*: AI systems have intelligence by using machine learning and deep learning technologies to understand the environment through data and arrive at a probabilistic conclusion. An example is performing the root-cause analysis (RCA) of an issue. These systems try to mimic human intelligence and decision-making, but technologically we are still very far from reaching levels of human intelligence in AI systems. However, in some specific tasks such as natural language understanding, some AI systems are providing accuracy levels that are on par with average human accuracy.

- *Adaptability*: AI systems have the ability to learn and adapt as they compile information and make decisions. AI systems have the ability to adapt as data changes. Thus, rather than learning from static rules, an AI system is highly adaptable and learns from changing data. For example, predicting the outage or service impact in business involves the ability to adapt to changing environments and new data.

# Types of Machine Learning

AI is quite an extensive and complex domain that consists of multiple specialized subdomains. Let's understand machine learning and deep learning in greater detail.

# Machine Learning

Machine learning is a technique that includes the algorithms that parse data, learn from the data supplied to the algorithm, and then apply the learning to make decisions. Unlike traditional programming where a person must manually create programs and rules based on the point-in-time snapshot of the environment and specific logic derived by the programmer, ML algorithms automatically derive that logic from data and create required rules.

Machine learning algorithms are at work in most of the applications that we use today. A simple example is a recommender system that recommends a movie, a song, or an item for us to buy based on analysis of our past preferences as well as the preferences of other users who are statistically similar to us.

As the name suggests, machine learning algorithms are not static; they continue to learn with new data that is being fed into the system, and they get progressively better at the job with more data.

The term *machine learning* was first coined by Arthur Samuel in 1952; its canonical and more practical definition was given by Tom Mitchell in 1997, which says "An agent or a program is said to learn from experience E with respect to some class of tasks T and performance measure P, if learner performance at tasks T, as measured by P, improves with experience E."

Machine learning has three different types, namely, supervised learning, unsupervised learning, and reinforcement learning, as shown in Figure 5-1.

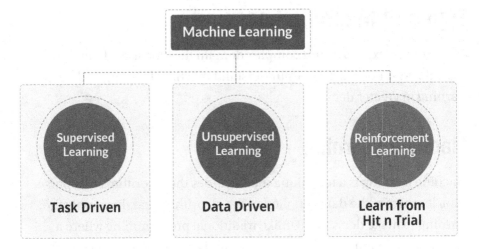

***Figure 5-1.*** *Types of machine learning*

# Supervised (Inductive) Learning

In supervised learning, users provide training data for a specific task to ML algorithms to analyze and learn from it. Training data includes both input, which is termed as *features*, and correct output, called the *label*. A simple example is an event with its description, timestamp, number of times it has occurred, the device on which it has occurred, and whether it is a critical alert or noise. Once the system gets enough data examples, it learns what is an alert and what is noise. The more training data, the higher the accuracy of the system.

This model is then applied to unseen data as an input for the algorithm to predict a response. The model is essential for the algorithm finding the connections between the parameters provided and establishing a cause-and-effect relationship between the features in the dataset. At the end of the training, the algorithm develops a draft of how the data works and the relationship between the input and the output.

For machine learning algorithms, the labeled dataset is divided into training data and test data where the training dataset trains the algorithm, and later the testing dataset is used to feed into the model to measure its accuracy of predictions. Models are then further fine-tuned progressively to arrive at better accuracy.

Both the quantity and quality of training data are important in determining the accuracy of output. As per best practices, the dataset should be split into a 70 to 30 ratio between training and testing datasets. If there is uneven class distribution in the dataset, then it is called an *imbalanced* training dataset, which means many more records for a particular class/label and very few records for another class/label. An imbalanced dataset causes imperfect decision boundaries, leading to inaccuracies in prediction. From an AIOps perspective, the training dataset should represent the actual production or operational characteristics with correct labeled parameters.

Another important factor in improving the accuracy of prediction is to remove irrelevant features as they negatively impact the learning process. The features are various inputs that are part of the data; it is important that only relevant features are provided in the training. Incorrect, insufficient, or irrelevant data in training will lead to errors in the model.

From an AIOps perspective, most commonly used supervised machine algorithms are as follows:

- *Regression*: Regression is used to predict a continuous numerical value. For instance, at any specific time interval, it finds out what is the percentage utilization of CPU or count of website hits, capacity (GB) of a database, etc.

- *Classification*: Classification is used if the expectation is to get a discrete value from a set of finite categories. We have already seen an example of classifying an event as an alert or noise. Classification use cases are marked as binary classification problem or multiclass classification problems depending upon whether the predicted output can belong to either of two classes or one of the multiple classes.

The following are supervised machine learning algorithms:

- Regression

- Logistic regression

- Classification

- Naive Bayes classifiers

- K-NN (k-nearest neighbors)

- Decision trees

- Support vector machine

With an understanding of supervised ML techniques, let's explore another crucial ML technique, that of unsupervised learning.

# Unsupervised Learning

One of the big challenges in AIOps implementation in any organization is to get clean labeled data. In such scenarios, unsupervised machine learning provides the starting point for AIOP's journey. Unlike supervised learning, it doesn't need any "correct answer" as a label. Rather, these algorithms explore inherent patterns and relationships from millions of unlabeled data points to discover hidden correlation and patterns in the data. The unsupervised learning algorithms can adapt to the data by dynamically changing hidden correlations. It allows the model to work on its own to discover patterns and information that were previously undetected. Since there is no training data and it is left to the algorithm to decipher and make sense of the data, it is computationally more intensive than supervised learning. Since there is no training provided and no expert input to the algorithm, it is also less accurate than supervised learning. At times unsupervised learning is used to arrive at an initial analysis, and then the data is fed to supervised learning algorithms after gaining more insights. Though unsupervised learning does not use labels, humans still need to analyze the output generated to make sense of the data and fine-tune the models so that the expected result is generated.

There are unsupervised machine learning algorithms that are mostly used in AIOps.

*Clustering* identifies the anomalies in the dataset by splitting, and clustering is based on the similarity and differences between features and patterns available in the dataset. Multiple clusters help the operations team to diagnose issues and anomalies. Clusters divide the entire data set into subsets that are more similar to each other, and thus it creates insights into the data without any training. Clustering will bring together similar events, and this can then be used to further analyze the data and select appropriate mechanisms and algorithms that need to be used for AIOps.

*Association* discovers the relationship between entities based on the correlation deduced from the dataset. For example, a long-running database job on a server can be associated with high CPU utilization and may lead to high response time on website hits or transactions.

From an AIOps perspective, datapoints should be exposed to unsupervised learning to leverage clustering and association algorithms for root-cause analysis purposes. As a best practice, all events and incidents should be fed into the unsupervised learning algorithm to determine noise.

Dimensionality reduction is a learning technique used when the number of features in a given dataset is too high. It reduces the number of data inputs to a manageable size while also preserving the data integrity. Often, this technique is used in the preprocessing data stage. This ensures that noisy features or features that are not relevant to the task at hand are reduced.

The following are types of unsupervised learning:

- Clustering

- Exclusive (partitioning)

- Agglomerative

- Overlapping

- Probabilistic

The following are the most commonly used clustering algorithms:

- *Hierarchical clustering*: As the name suggest, this
  clustering algorithm uses a "hierarchy" of clusters to
  decompose data into a cluster and form a tree structure
  called a *dendrogram*. From an AIOps perspective, it
  is useful in automatically detecting service models
  without a CMDB and performing service impact
  analysis on it. You can either follow a top-down
  approach (divisive) or a follow bottom-up approach
  (agglomerative) to group data points in a cluster. In
  the divisive approach, all data points are assumed to
  be part of one large cluster (like application service),
  and then based on the termination logic they get
  divided into smaller clusters (like technical service).
  In agglomerative, each data point is assumed to be a
  cluster, which iteratively gets merged to create large
  clusters.

- *Centroid-based clustering*: These algorithms are one
  of the simplest and most effective ways of creating
  clusters and assigning data points to them. In this
  algorithm, we have to first find "K" centroids and then
  group dataset based on the distance (or proximity)
  with centroids. K-means is a popular centroid-based

clustering algorithm, and we will be exploring it in detail in Chapter 8 for anomaly detection, noise detection, and probable cause analysis processes as part of the AIOps implementation in IT operations.

- *Density-based clustering*: Both centroid-based and hierarchical clustering techniques are based on the distance (or proximity) between data points without considering the density of data points at a specific position (or time stamp). This limitation is handled by a density-based clustering algorithm, and the techniques perform much better than K-means algorithms for outlier detection and noise from data set.

Next, we will discuss reinforcement learning, which is another technique that is used in specific situations.

# Reinforcement Learning

Both supervised and unsupervised algorithms have a dependency on the dataset for learning. But you can also learn from feedback on your actions whether it is in the form of a reward or penalty. Reinforcement learning leverages this methodology and is driven by feedback from the sequence of action. It continuously improves and learns by the trial-and-error method. The reinforcement learning algorithms are based on training that involves reward and penalty. Thus, for every successful action, a reward is given to the agent, and for every unsuccessful step a penalty is levied. The reinforcement learning policy determines what is expected from the agent. It also involves tweaking the short-term and long-term rewards and penalties so that the agent does not go for short-term rewards while losing out on a bigger reward in the longer run. Apart from the agent, reward and penalty, and policy, the other important element is the environment in which the agent is operating, so at every step of the agent the state of the agent and its environment is fed back into the system to calculate what steps to take next. Reinforcement learning powers some of the AI-based systems that have outplayed human agents in traditional and video games.

Figure 5-2 summarizes the algorithms that we have discussed in machine learning.

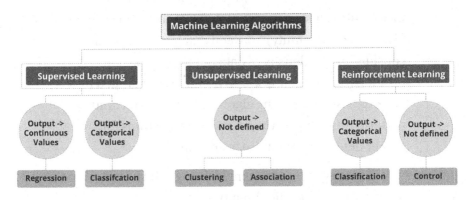

***Figure 5-2.***  *Types of ML algorithms*

# Differences Between Supervised and Unsupervised Learning

In supervised learning, the goal is to predict outcomes for new data by leveraging the learnings from the training data. In this model, the categories are known for the result, so you know what the end result will be from a set of possible outcomes. With an unsupervised learning algorithm, the goal is to get insights from the available data, and the algorithm decides on the features from the data provided. The output of unsupervised learning and their categories is not known.

Supervised learning models are ideal for data where we know the categories and we have sufficient training data available to train the machine learning algorithm. Supervised learning algorithms are used extensively in price prediction, forecasting, sentiment analysis, and other classification and regression tasks. Unsupervised machine learning is used extensively in recommendation engines, anomaly detection, etc.

Supervised leaning algorithms are more accurate since the training data is curated, validated, and provided by subject-matter experts, while unsupervised algorithms are not as accurate and require human intervention for interpretation of the result. Training supervised learning algorithms requires time and effort to create the training data, while for unsupervised learning, there is no requirement to create training data.

With an understanding of all three learning techniques, let's understand how to choose from various options that are available to us for implementing these technologies.

# Choosing the Machine Learning Approach

Choosing the right type of machine learning approach depends on various factors.

- *Objective*: Is the objective or goal to understand the data and its correlation and features in more detail? Is the goal to cluster together similar data for analysis? Or is the goal a probabilistic prediction of a discrete or continuous variable? Dependent on the end goal of the application, you need to choose between supervised and unsupervised machine learning.

- *Data*: Is there availability of training and labeled data? Can this data be made available? If yes, then you can go for supervised learning; otherwise, in absence of training data, you need to opt for unsupervised machine learning techniques.

ML techniques discussed so far in AIOps may need to deal with a lot of textual data and not just the metrices. Natural language processing (NLP) made it possible to analyze text data using ML techniques and will be discussed next.

# Natural Language Processing

NLP is one of the key research fields within the AI domain that deals with processing information contained (or rather hidden) in natural language text. NLP uses language semantics and syntax to determine the context as well as real meaning and emotions getting conveyed via text.

NLP enables an AIOps system to mine knowledge from various rulebooks and knowledge articles available within the organization as well as search the latest information available in vendor repositories or online communities. It is practically impossible for an individual to scan and ingest knowledge from various thousands of documents available and accordingly perform specific tasks or actions in a limited time window. Knowledge obtained from NLP helps tremendously in improving reinforcement learning algorithms, executing automated resolutions efficiently, and providing recommendations and guidance to SRE and DevOps while dealing with an outage.

Chatbots, which were once considered as an optional entity, have become a must-have for business, especially in the service industry. Thanks to the emerging technology, self-service capabilities are given to end users, which makes them feel empowered. With NLP, chatbots can have a human-like meaningful conversation instead of just giving predefined limited responses to queries. This becomes a huge differentiator in an AIOps system. Let's understand the natural language process in more detail.

# What Is Natural Language Processing?

Natural human language is complex; it has various variations that may mean the same thing. It is ambiguous at times, it is dependent on the context of the object or situation, and it is extremely diverse. Thus, it comes with its own set of challenges when we try to make machines interpret or decipher the language. We haven't used the word *understand* here, because the machines may not have an understanding of the language like we have. That's why deciphering and interpreting what we are trying to communicate is more appropriate.

NLP is the domain in artificial intelligence that makes it possible for computers to understand human language. NLP analyzes the grammatical structure of sentences and the individual meaning of words; it uses machine learning and deep learning algorithms and other specific NLP techniques to extract meaning from the provided input. Thus, NLP is the technology that makes machines decipher human language, and machines can take input using natural human language rather than software code.

NLP technology is used extensively in various use cases; however, it is most visible in the form of virtual assistants like Apple Siri, Google Assistant, Microsoft Cortana, and Amazon Alexa. You will also find cognitive virtual assistants in many applications and websites where you can type in your query in human language and the system will be able to decipher it and return an answer. All these systems use NLP technologies behind the scenes.

Beyond the chatbots and virtual assistants, there are many other applications that use NLP. Text recommendations or next word suggestions when you are trying to key in terms in a search engine, language translation when you use Google Translation Services, spam filtering in email, sentiment analysis of posts or Twitter feeds—all these use cases use NLP technologies.

In a nutshell, the goal of NLP is to make human language—which is complex, ambiguous, and extremely diverse—easy for machines to understand.

NLP and machine learning are both subsets of artificial intelligence. To create NLP algorithms, machine learning technologies are used. Since it is a domain in itself, NLP is called out separately as a domain in artificial intelligence technologies. There are techniques, algorithms, and systems entirely devoted to NLP.

NLP applies two techniques to help computers understand text: syntactic analysis and semantic analysis.

# Syntactic Analysis

Syntactic analysis—or parsing—analyzes text using basic grammar rules to identify sentence structure, how words are organized, and how words relate to each other. Some of its main subtasks include the following:

- *Tokenization* consists of breaking up text into smaller parts called tokens to make text easier to handle. This is the first step in NLP to break down the text into tokens so that it can be processed further by the NLP engine.

- *Part-of-speech tagging (PoS tagging)* labels the tokens that are generated from tokenization as noun, adjective, verb, adverb, etc. This helps the NLP engine to infer the meaning of the word in a particular context. For example, the word Saw can mean seeing in the past or it can be a noun pointing to the object Saw.

- *Stop-word removal* removes frequently occurring words that don't add any semantic value, such as *I, they, have, like, yours*, etc.

- *Stemming* converts a word into root form by removing suffixes. For example, *Studying* will be converted to *Study*. It refers to a crude heuristic process that trims ends of words to find the correct root most of the time but is less resource extensive as well as fast.

- *Lemmatization* also converts the word into the root but by using vocabulary and morphological analysis of word with the aim to remove inflectional endings and return the base or dictionary form of a word. For example, it replaces an inflected word with its base form, so *Saw* gets converted into *See*. Lemmatization is more accurate but more resource extensive and complex.

- *Anaphora resolution* deals with the problem of resolving what a pronoun such as *he* or a noun such as *CEO* refer to.

# Semantic Analysis

Semantic analysis focuses on capturing the meaning of text. It follows a two-step process. First it tries to extract the meaning of each individual word, and then it looks at the combination of words to decipher what they mean in a particular context.

- *Word-sense disambiguation*: This deals with the problem of determining the true meaning or sense of a word that is used in the sentence. This impacts the output of anaphora resolution as well.

- *Relationship extraction*: Relationship extraction analyzes the textual data and tries to find relationships between various entities. Entities can be various nouns such as people, places, geographies, objects, etc.

NLP tools and techniques are highly applicable in the AIOps domain. Chatbots or cognitive virtual agents are one of the modules in AIOps powered by NLP technologies. Within the cognitive virtual assistant there are multiple NLP services and techniques that are combined to deliver the cognitive virtual assistant. With an understanding of the NLP technique, let's discuss its AIOps use cases.

# NLP AIOps Use Cases

NLP plays a crucial role in AIOps systems in the understanding of textual data in alert messages (SNMP traps, emails, event IDs, etc.), incidents, user chat messages, log contents, and many other sources that convey issues or feedbacks. NLP enables the interpretation of such text for AIOps systems and accordingly takes the appropriate actions. Let's explore some of the NLP use cases in AIOps.

## Sentiment Analysis

Sentiment analysis identifies emotions in text and classifies the data as positive, negative, or neutral.

While using ticket data or chat data, sentiment analysis is a key use case. For every ticket feedback and for every conversation initiated and closed with the agents, you can leverage sentiment analysis to find out how the users feel about the services rendered. This comes in handy during a cognitive virtual assistant conversation as well because the cognitive virtual agent can tailor its responses based on the sentiment of the user. As an example, if a user is angry, the cognitive virtual agent can start the conversation with a pacifying statement like "sorry for the inconvenience caused" and make the conversation more natural and human-like rather than a robotic conversation where the sentiment may get ignored.

# Language Translation

Many IT operations support multiple geographies and languages, and it becomes difficult for IT service administrators and service desk agents to cover 24/7 for all languages. Thus, language translation services come in handy for service desk agents to translate the ticket data or conversation data and use that to pinpoint the user's issue and resolve it. Language translation is used both in cognitive virtual assistants and in user-to-agent communication. Translation services also find their utility in analyzing the ticket data that may have been provided in different languages.

# Text Extraction

Text extraction enables you to pull out predefined information from text. IT teams typically deal with large volumes of data in various knowledge management repositories. There is also data in known error databases (KEDBs) in IT service management systems. There is also voluminous data in technical documentation that provides information for root-cause analysis, troubleshooting, and resolution. All this information becomes overwhelming for operations teams. AIOps systems extract the relevant information from these repositories and present the right information to administrators and service desk agents so they can troubleshoot and resolve issues or complete service request and changes.

# Topic Classification

Topic classification helps the AIOps engine to organize the unstructured text available in various repositories into categories. Most of the time information is spread across multiple systems, and any directory structure or categorization structure that was initially created becomes unmanageable. Content for related categories and topics gets spread out in various repositories, and manually maintaining the topics and relevant documents becomes impossible. The AIOps engine helps you to organize this unstructured data into various categories. This feature can also be used to tag the tickets into various categories in IT service management.

This covers the overview of the NLP technique and its use case, and now we will move to the last but most complex technique that is used in AIOps, which is deep learning.

# Deep Learning

We covered machine learning and its branches in the previous sections. As introduced earlier, deep learning is a subset of machine learning or a specialized way of making machines learn from data. Deep learning functions in a similar fashion as machine learning; however, the technologies used to achieve the end results and the capabilities that deep learning provides are a bit different.

Though both machine learning and deep learning are used interchangeably, both are different technically in their approach to learning.

Deep learning is a subfield of machine learning that structures algorithms in layers to create an "artificial neural network" that can learn and make intelligent decisions on its own.

A deep learning model is designed to analyze data in a structure similar to how humans draw conclusions. Deep learning algorithms use a layered structure called an *artificial neural network* to process the data and learn various features in the data.

Artificial neural networks have been in existence theoretically for a long time; however, the application of artificial neural networks was possible only once the computing technology progressed to provide the compute resources necessary to run these models. Thus, neural networks and deep learning became practical and created miraculous solutions in the last decade. Google created AlphaGo, which beat all human champions at the game, and this demonstrated the arrival of computing intelligence, which could be applied to games that required billions of computations to solve.

Artificial neural networks were inspired by biological neural networks in animal and human brains. The biological neural networks learn progressively by seeing and learning from data without any task-specific program guiding them. A human is able to recognize a cat from a dog after just seeing a few samples of cats and dogs; however, a machine would need much more data to learn, but it surely learns the difference and is able to classify them under the right species after learning.

Artificial neural networks are created using layers of neurons. Different layers called *hidden layers* perform a different transformation on the input as the signal moves from the input layer through the various hidden layers to the output layer, which finally arrives at the result as shown in Figure 5-3.

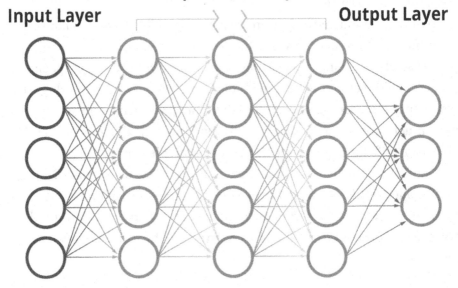

*Figure 5-3.* *Deep learning neural network*

Neural networks can have a huge number of neurons and connections spanning millions of connections for complex use cases.

Neural networks and deep learning are being used in various use cases such as NLP, machine translation, image recognition, video gameplays, spam filtering, search engines, etc.

Neural networks at a fundamental level comprise inputs, weights, a bias (threshold), and an output. The neural network then balances the weight when training data is fed into the system using backpropagation to find out which neurons cause the errors in mapping to the correct output. After the neural network is trained with all the data in multiple iterations, the weights of each neuron get configured in a manner that the collective response as an output is accurate.

Threshold is an important parameter in neural networks; it leads to the activation of the node if the value generated is above the threshold, and it then sends the data to the next layer. Thus, the output of one neuron in a layer impacts the input and thus the output of the next layer of neurons to which it is connected. The hidden layers comprising multiple neurons have their own activation functions and keep passing the information based on the threshold to the next layer.

The idea is that each layer discovers a feature in the input, and progressively lower-level features are discovered by the next layer, just like a human brain would work, finding a higher-level object and then discovering what the object is and its features.

The general equation for a neural network is as follows, where $w$ is the weight and $x$ are the features:

$$\sum_{i=1}^{m} w_i x_i + bias = w_1 x_1 + w_2 x_2 + w_3 x_3 + bias$$

Neutral networks use backpropagation, which allows the weights in neurons to adjust in response to the error that they are contributing to the output. This allows for the adjustment of the weights and the model to progressively reach higher levels of accuracy and converge so that it can then be deployed for the use case for which it is getting trained.

Though a deep learning model needs much higher computing resources and data, the flexibility of the model and its ability to learn complex features makes it a compelling proposition for various use cases. Deep learning is generally deployed in use cases where there is higher complexity, the amount of data is much higher, and there is availability of data for training the deep learning model.

# Summary

This chapter covered the basics of machine learning, deep learning, natural language processing, and how some of the capabilities can be used in the AIOps domain. We covered various techniques including supervised and unsupervised machine learning as well as NLP. These techniques provide the foundation on which AIOps platforms are built. In the next chapter, we will look at specific AIOps use cases and how we can use these techniques in those cases. We will start with the most common use case of deduplication in the next chapter.

# CHAPTER 6

# AIOps Use Case: Deduplication

In the previous chapters, you learned about the various machine learning techniques and how they are used in general. In this chapter, you will see the implementation of a specific use case of deduplication in AIOps along with practical scenarios where it is used in operations and business.

## Environment Setup

In this section, you'll learn how to set up the environment to run the sample code demonstrating various AIOps use cases. You have three options to set up an environment.

- Download Python from `https://www.python.org/downloads/` and install Python directly with the installer. Other packages need to be installed explicitly on top of Python.

- Use Anaconda, which is a Python distribution and made for large data processing, predictive analytics, and scientific computing requirements. The Anaconda distribution is the easiest way to perform Python coding, and it works on Linux, Windows, and macOS. It can be downloaded from `https://www.anaconda.com/distribution/`.

© Navin Sabharwal and Gaurav Bhardwaj 2022
N. Sabharwal and G. Bhardwaj, *Hands-on AIOps*,
https://doi.org/10.1007/978-1-4842-8267-0_6

- Use cloud services. This is the simplest of all the options but needs Internet connectivity to use. Cloud providers such as Microsoft Azure Notebooks and Google Collaboratory are popular and available at the following links:

  - Microsoft Azure Notebooks: `https://notebooks.azure.com/`

  - Google Collaboratory: `https://colab.research.google.com`

We will be using the second option and will set up an environment using the Anaconda distribution.

# Software Installation

Download the Anaconda Python distribution from `https://www.anaconda.com/distribution/`, as shown in Figure 6-1.

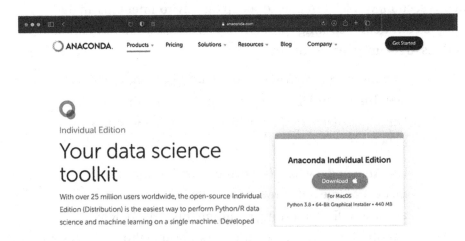

*Figure 6-1.  Downloading Anaconda*

After downloading the file, trigger the setup to initiate the installation process.

1. Click Continue.

2. On the Read me step, click Continue.

3. Accept the agreement, and click Continue.

4. Select the installation location, and click Continue.

5. It will take about 500MB of space on the computer. Click Install.

6. Once the Anaconda application gets installed, click Close and proceed to the next step to launch the application.

# Launch Application

After installing Anaconda Jupyter, open the command-line terminal and type `jupyter notebook`.

This will launch the Jupyter server listening on port 8888. Usually, a pop-up window with the default browser will automatically open, or you can log in to the application by opening any web browser and opening the URL `http://localhost:8888/`, as shown in Figure 6-2.

Click New ➤ Python 3 to create a blank notebook.

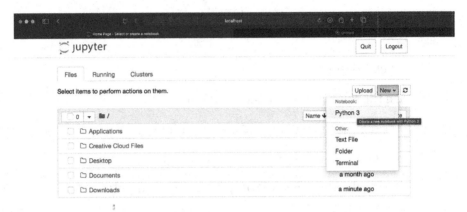

**Figure 6-2.** *Jupyter Notebook UI*

A blank notebook will launch in a new window. You can just type commands in the cell and click the Run button to execute them. To test that the environment is working perfectly, type in a print ("Hello World") statement in the cell and execute it.

Python Notebooks are used in the book for demonstrating the AIOps use cases of deduplication, dynamic baselining, and anomaly detection; they are available at `https://github.com/dryice-devops/AIOps.git`. You can download the source code to follow along with the exercises in this chapter.

The environment is now ready to develop and test algorithmic models for multiple AIOps use cases. Before getting into model development, let's understand a few terms that are going to be used to analyze and measure the performance of models.

# Performance Analysis of Models

In practice, multiple models need to be created to make predictions, and it is important to quantify the accuracy of their prediction so that the best-performing model can be selected for production use.

In the supervised ML model, a set of actual data points is separated into training data and test data. The model is trained to learn from training data and make predictions that get validated against both training data and testing data. The delta between actual values and predicted values is called the *error*. The objective is to select a model with minimum error.

There are multiple methods to analyze errors and measure the accuracy of predictions. The following are the most commonly used methods.

# Mean Square Error/Root Mean Square Error

In this approach, the first error is calculated for each data point by subtracting the actual result value from the predicted result value. These errors need to be added together to calculate the total error obtained from a specific model. But errors can be positive or negative on different data points, and adding them will cancel out each other, giving an incorrect result. To overcome this issue, in mean square error (MSE), a square of calculated error is performed to ensure that the positive and negative error values should not cancel out each other, and then the mean of all squared error values is performed to determine the overall accuracy.

$$\text{MSE} = \frac{1}{n} \sum_{i=1}^{n} (Y_i - \hat{Y}_i)^2.$$

$n$ -> Number of data points

$Y_i$ -> Actual value at specific test data point

$\hat{Y}_i$ -> Predicted value at specific test data point

There is another variant of MSE where the square root of MSE is performed to provide the root mean square error (RMSE) value. The lower the MSE/RMSE, the higher the accuracy in prediction and the better the model.

$$RMSE = \sqrt{\sum_{i=1}^{n} \frac{(\hat{y}_i - y_i)^2}{n}}$$

Both MSE/RMSE use "mean" as the baseline, and that's the biggest issue with these methods because mean has inherent sensitivity toward outliers. In simple terms, assume there are four data points provided to model A and model B. Table 6-1 lists the errors obtained on each of the data points for each model.

***Table 6-1.*** *Sample Data and Errors in ML Models*

| Data Point | Error in Model A | Error in Model B |
| --- | --- | --- |
| D1 | 0.45 | 0 |
| D2 | 0.27 | 0 |
| D3 | 0.6 | 0 |
| D4 (Outlier) | 0.95 | 1.5 |

Model B appears to be a better model because it predicted with 100 percent accuracy except on one data point, which was an outlier and gave a high error. On the other hand, model A gave small errors D on almost every data point.

But if we consider the mean of errors for both models, then model A will have a lower mean value of errors than model B, which gives an incorrect interpretation of model accuracy.

# Mean Absolute Error

Similar to MSE, in the mean absolute error (MAE) approach the first error is calculated as the difference between the actual result value and the predicted result value, but rather than squaring the error value, MAE considers the absolute error value. Finally, the mean of all absolute error values is performed to determine the overall accuracy.

$$MAE = \frac{1}{n}\sum_{i=1}^{n}\left|Y_i - \hat{Y}_i\right|$$

# Mean Absolute Percentage Error

One of the biggest challenges of MAE, MSE, and RMSE is that they provide absolute values that can be used to compare multiple models' accuracies. But there is no way to validate the effectiveness of an individual model as there is no scale defined to compare that absolute value. The output of these methods can take any value, and there is no range defined.

Instead of calculating the absolute error, MAPE measures accuracy as a percentage by first calculating the error value divided by the actual value for each data point and then taking the mean of them.

$$MAPE = \frac{100\%}{n}\sum_{i=1}^{n}\left|\frac{y_i - \hat{y}_i}{y_i}\right|$$

MAPE is a useful metric and one of the most commonly used methods to calculate accuracy because the output lies between 0 and 100, enabling intuitive analysis of the model's performance. It works best if there are no extremes or outliers and no zeros in data point values.

# Root Mean Squared Log Error

One of the challenges in RMSE was the effect of outliers on the overall RMSE output value. To minimize (sometimes nullify as well) the effect of these outliers, there is another method called RMSLE, which is a variant of RMSE and uses the logarithmic property to calculate the relative error between the predicted value and the actual values instead of the absolute error.

$$RMSLE = \sqrt{\frac{1}{n} \sum_{i=1}^{n} (log(\hat{y}_i + 1) - log(y_i + 1))^2}$$

In cases when there is a huge difference between the actual and predicted values, the error becomes large, resulting in huge penalties by RMSE but handled well by RMSLE. Also, to note that in cases where either the predicted value or actual value is zero, then the log of zero becomes not defined. That's why one is added to both the predicted and actual values to avoid such situations.

# Coefficient of Determination-R2 Score

In statistics, R2 is defined as the "proportion of the variation in the dependent variable that is predictable from the independent variable(s)." It measures the strength of the relationship between the model and the dependent variable on a convenient 0–100 percent scale.

- 0 percent represents a model that does not explain any of the variations in the response variable around its mean.

- 100 percent represents a model that explains all the variation in the response variable around its mean.

After fitting a linear regression model, it is important to determine how well the model fits the data. Usually, the larger the R2, the better the regression model fits your observations; however, there are caveats that are beyond the scope of this book.

Having understood the various ways to measure accuracy of models, we will now proceed with our first AIOps use case, which is a deduplication of events.

# Deduplication

Deduplication is one of the most primitive features of the event management function to reduce noise by processing millions of events arriving from multiple tools that are monitoring the underlying infrastructure and applications. Consider a simple and common scenario where a batch processing job gets triggered causing CPU utilization to cross the warning threshold of 70 percent and within 15 minutes backup gets initiated, shooting up the CPU utilization to 90 percent and generating a critical event. This event will be repeated for every poll that the monitoring system makes to this system, and once the scheduled job is over, the CPU utilization will return to its normal state. However, within this period, it has generated four events with different timestamps. But essentially, all these events represent the same issue of high CPU utilization on a server. There is a plethora of such operational scenarios, both simple and complex, that generate duplicate events. As shown in Figure 6-3, the deduplication function manages this complexity to validate the uniqueness of a new event by comparing it against an existing event in the system, dropping it if a match is found along with an increasing counter of the original alert, and updating the timestamp based on the latest event generated. This helps in ensuring that the event console is not cluttered with the same event multiple times.

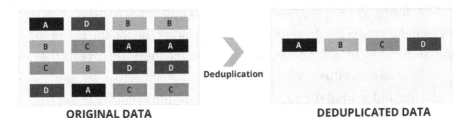

*Figure 6-3.* *Deduplication function*

To implement the deduplication function, a unique key needs to be defined that defines unique issues and their context. This unique key gets created dynamically using information coming in the event such as the hostname, source, service, location, class, etc. A rule or algorithm can be configured to match this unique key with all events in the system and perform the deduplication function. Please note that this AIOps function does not need machine learning and can be simply done by using a rule-based deduplication system.

Let's observe the practical implementation of the deduplication function. This implementation will be reading alerts from an input file.

In our code, we will be using Pandas, which is a powerful library of Python to analyze and process alerts, along with the SQLite DB, which is a lightweight disk-based database to store alerts after processing. Unlike client–server database management systems, SQLite doesn't require a separate installation for a server process and allows accessing the database using a nonstandard variant of the SQL query language. SQLite is popular as an embedded database for local/client storage in software such as web browsers. SQLite stores the entire database (definitions, tables, indices, and the data itself) as a single cross-platform file on a host machine.

After reading alerts from an input file, a unique key of EMUniqueKey will be dynamically created to determine the uniqueness of the issue and execute the deduplication function. At the end of the code, we will determine how many alerts were genuine alerts in our code and how many were duplicates.

In this code we will be using the Python libraries mentioned in Table 6-2.

**Table 6-2.** *Sample Data and Errors in ML*

| Library Name | Purpose |
|---|---|
| matplotlib | It is a data visualization library containing various functions to create charts and graphs for analysis. It supports multiple platforms and hence works on many operating systems and graphics backends. |
| Pandas | It is extensively used for data manipulation tasks. It is built on top of two core Python libraries—matplotlib for data visualization and NumPy for performing mathematical operations. |
| sqlite3 | This library provides an API 2.0 interface for SQLite databases. |

Open a new Jupyter notebook or load the already provided notebook from the GitHub download. Download the notebook from `https://github.com/dryice-devops/AIOps/blob/main/Ch-6_Deduplication.ipynb`.

Let's start by importing the required libraries.

```
import pandas as pd
import sqlite3
from matplotlib import pyplot as plt
```

Now read data from file and perform descriptive analysis to understand it.

```
raw_events = pd.read_excel("Alert-Bank.xlsx")
```

Let's find how many rows and columns we have in a dataset.

```
raw_events.shape
```

In the output, as shown in Figure 6-4, we have 1,051 total sample events in the alert bank, and each event has nine columns as an event slot.

Out[2]: (1051, 9)

*Figure 6-4. Count of data points in input file*

Let's list all columns with their data types and the number of non-null values in each column.

raw_events.info()

As per the output shown in Figure 6-5, there are 1,051 non-null values, which mean that there are no null values in any column. In cases where the column contains null values, then it needs to be processed before proceeding further. Usually, if the count of null values is less, then the rows containing null values can be removed completely, or else null values can be replaced with mean or median values.

```
<class 'pandas.core.frame.DataFrame'>
RangeIndex: 1051 entries, 0 to 1050
Data columns (total 9 columns):
 #   Column            Non-Null Count  Dtype
---  ------            --------------  -----
 0   Host              1051 non-null   object
 1   AlertTime         1051 non-null   datetime64[ns]
 2   AlertDescription  1051 non-null   object
 3   AlertClass        1051 non-null   object
 4   AlertType         1051 non-null   object
 5   AlertManager      1051 non-null   object
 6   Source            1051 non-null   object
 7   Status            1051 non-null   int64
 8   Severity          1051 non-null   int64
dtypes: datetime64[ns](1), int64(2), object(6)
memory usage: 74.0+ KB
```

*Figure 6-5. Deduplication function*

Next, it is important to understand the values contained in these slots, and for that, we execute the head() command to check the values of the first five rows along with transpose() for better understanding and visibility of data. In the output snippet, shown in Table 6-3, we can see all five fields of three events from the alert bank.

```
raw_events.head().transpose()
```

***Table 6-3.*** *Output Snippet Showing Sample Events*

|  | 0 | 1 | 2 | 3 |
|---|---|---|---|---|
| **Host** | AUPRDGLB-HRAPP03 | USPRDEMPRECWEB03 | USDEVGLB-HRAPP02 | UKDEVPAYROLDBA01 |
| **AlertTime** | 2021-06-02 03:56:34.995000 | 2021-06-02 05:55:32 | 2021-06-02 07:50:33.995000 | 2021-06-02 07:50:47.995000 |
| **AlertDescription** | Memory Utilization is 74% in Warning State on... | Memory Utilization is 70% in Warning State on... | CPU Utilization is 95% in Critical State on U... | CPU Utilization is 93% in Critical State on U... |
| **AlertClass** | OperatingSystem | OperatingSystem | OperatingSystem | OperatingSystem |
| **AlertType** | Server | Server | Server | Server |
| **AlertManager** | PlatformMonitoring-Tool | PlatformMonitoring-Tool | PlatformMonitoring-Tool | PlatformMonitoring-Tool |
| **Source** | Memory | Memory | CPU | CPU |
| **Status** | 3 | 1 | 1 | 1 |
| **Severity** | 2 | 2 | 1 | 1 |

Based on the output, we can observe that each event contains the following details:

- *Host*: Mentions CI from where the event got generated.

- *AlertTime*: Mentions the timestamp of the event when it got generated at CI.

- *AlertDescription*: Mentions details about issues for which event got generated.

- *AlertClass*: Mentions the class to which the event belongs such as the operating system.

- *AlertType*: Mentions the type to which the event belongs such as the server.

- *AlertManager*: Mentions the manager that captured the issue and generated the event. It's usually the underlying monitoring tools such as Nagios, Zabbix, etc.

- *Source*: Mentions the source for which the issue was detected such as the CPU, memory, etc.

- *Status*: Mentions the event status, which is a numerical value.

- *Severity*: Mentions the event severity, which is again a numerical value.

For the purpose of this example, we have assigned a unique code to different values in the Status field as well as the Severity field. There can be different values assigned to Event Status, as shown in Table 6-4, as well as different values assigned to Event Severity, as shown in Table 6-5. These mappings varied a lot in different tools.

***Table 6-4.*** *Event Status Code Description*

| Status Value | Meaning |
| --- | --- |
| 1 | Open state of event |
| 2 | Acknowledge state of event |
| 3 | Resolved state of event |

**Table 6-5.** *Event Severity Code Description*

| Severity Value | Meaning |
| --- | --- |
| 1 | Critical severity of event |
| 2 | Warning severity of event |
| 3 | Information severity of event |

Let's list the unique values of the Status columns in a dataset.

```
raw_events['Status'].unique()
```

As per the output in Figure 6-6, there are three types of event severity present in the input file.

```
Out[9]:  array([3, 1, 2])
```

**Figure 6-6.** *Deduplication function*

Based on the analysis done, we can conclude that each event has a specific AlertTime that got generated from the device mentioned in Host with an event source captured in Source that belongs to a specific category defined by AlertClass and AlertType. This event got captured by the monitoring tool mentioned in AlertManager with a specific severity Severity and Status as an enumerated value. Issue details are provided in AlertDescription.

For the purpose of the deduplication example, let's name the deduplication key EMUniqueKey and create it as a combination of the Host, Source, AlertType, AlertManager, and AlertClass fields concatenated with the separator ::, as shown in Figure 6-7. Then let's add it in our dataset for each event.

**Figure 6-7.** *Unique key composition*

```
raw_events['EMUniqueKey'] = raw_events['Host'].str.strip()
+ "::" \
                + raw_events['Source'].str.strip() + "::" \
                + raw_events['AlertType'].str.strip() + "::" \
                + raw_events['AlertManager'].str.strip()
                  + "::" \
                + raw_events['AlertClass'].str.strip()
raw_events['EMUniqueKey'] = raw_events['EMUniqueKey'].
str.lower()
raw_events = raw_events.sort_values(by="AlertTime")
raw_events["AlertTime"] = raw_events["AlertTime"].astype(str)
```

EMUniqueKey is a generic identifier that uniquely identifies the underlying issues to provide the required context for the purpose of performing deduplication.

Let's observe the data after adding the unique identifier in the dataset.

```
raw_events.head().transpose()
```

As observed in the output in Table 6-6, we have a new field added called EMUniqueKey in the dataset for each event.

***Table 6-6.*** *Updated Dataset After Adding a Unique Identifier*

Out[11]:

| | 0 | 1 | |
|---|---|---|---|
| Host | AUPRDGLB-HRAPP03 | USPRDEMPRECWEB03 | USDEVGLB |
| AlertTime | 2021-06-02 03:56:34.995 | 2021-06-02 05:55:32.000 | 2021-06-02 07 |
| AlertDescription | Memory Utilization is 74% in Warning State on... | Memory Utilization is 70% in Warning State on... | CPU Utilization is 95% in Cr |
| AlertClass | OperatingSystem | OperatingSystem | Opera |
| AlertType | Server | Server | |
| AlertManager | PlatformMonitoring-Tool | PlatformMonitoring-Tool | PlatformMoni |
| Source | Memory | Memory | |
| Status | 3 | 1 | |
| Severity | 2 | 2 | |
| EMUniqueKey | auprdglb-hrapp03::memory::server::platformmoni... | usprdemprecweb03::memory::server::platformmoni... | hrapp02::cpu::server::platforn |

Now we are all set to perform the deduplication function on the given dataset.

Before we proceed, it is important to note that we can use arrays as well in code to read and process events fields by referring them with indexes (0,1,2, etc.), but it will limit the scalability if new fields need to be added later. To avoid the scalability issue, we will be using a dictionary datatype so that we can use the event field name instead of indexes, as shown in the following function:

```
def transform_dictionary(cursor, row):
    key_value = {}
    for idx, col in enumerate(cursor.description):
        key_value[col[0]] = row[idx]
    return key_value
```

Now values will be stored as a key-value pair, and we can directly refer to any event field with its name.

Let's define a class to contain a few critical functions for SQLite. First create a connection with the database by using the initialization function _init.

```
class Database:
# create db connection to store and process events.
    def __init__(self):
        self._con = sqlite3.connect('de_duplicationv5.db')
        self._con.row_factory = transform_dictionary
        self.cur = self.con.cursor()
        self.createTable()
    def getConn(self):
        return self.con
```

The function getConn will create a connection with the database.

Next, create the required tables in database. Though we need only one table dedup to store the final deduplicated events, we will also be creating a table archive to store duplicated events. The table archive will be used to quantify the benefits of the deduplication function.

```
def createTable(self):
        self.cur.execute('''CREATE TABLE if not exists dedup
                (AlertTime date, EMUniqueKey text, Host text, \
                AlertDescription text, AlertClass text,
                AlertType text, \
                AlertManager text, Source text, Status text, \
                Severity text, Count real)''')
        self.cur.execute('''CREATE TABLE if not exists archive
                (AlertTime date, EMUniqueKey text, Host text,
                AlertDescription text, AlertClass text,
                AlertType text,\
                AlertManager text, Source text, Status text, \
                Severity text)''')
```

After reading events, they need to be inserted either into the dedup table or into the archive table depending upon whether it's a new event or a duplicate event. The function insert will accept the table name and event values that need to be inserted. Before inserting the event, EMUniqueKey is getting dynamically generated for that event.

```
def insert(self, table, values):
    columns = []
    val = []
    value = []
    for EMUniqueKey in values:
        columns.append(EMUniqueKey)
        val.append("?")
        value.append(values[EMUniqueKey])
    query = "Insert into {} ({}) values ({})".format\
(table, ",".join(columns), ",".join(val) )
    self.cur.execute(query, tuple(value))
```

We need the function execute to execute database queries as well as the function update to update the event's count, severity, and status in the dedup table for any repeated occurrence of an event that is already present in the system in an open state.

```
def execute(self, query):
    self.cur.execute(query)
def update(self, query, values):
    # print(query)
    return self.cur.execute(query, values)
```

Finally, we need supporting functions to fetch record(s), commit transactions, and close the connection.

```
def fetchOne(self, query):
    self.cur.execute(query)
    return self.cur.fetchone()
```

```
def fetchAll(self, query):
    self.cur.execute(query)
    return self.cur.fetchall()
def commit(self):
    self._con.commit()
def close(self):
    self._con.close()
```

Now let's start reading and processing events data iteratively from the event bank. In the AIOps solution, this processing happens as part of stream processing in real time rather than as batch processing.

```
db = Database()
for item in raw_events.iterrows():
    #read events
    data = dict(item[1])
    print("Input Data",data)
    dedupData = db.fetchOne("Select * from dedup where
EMUniqueKey='{}' \
                    and Status != 3".format(data["EMUniqueKey"]))
    if dedupData:
        #increase count and add current row in archive
        count = dedupData['Count'] + 1
        query = "Update dedup set Count=?,
        AlertDescription=?, \
                Severity=?, Status=? where EMUniqueKey=? and
                Status=?"
        db.update(query, (count, data['AlertDescription'], \
                        data["Severity"], data["Status"],
                        data["EMUniqueKey"],
                        dedupData['Status']) )
        db.insert("archive", data)
        db.commit()
```

```
else:
    #insert in dudup table
    data['count'] = 1
    db.insert("dedup", data)
    db.commit()
```

In the previous code, first EMUniqueKey got generated dynamically using event slots and stored in the data variable (data frame). Then the table dedup gets checked to find out whether there is any open event with the same EMUniqueKey. If there is any matching event found in the dedup table, then it means that a new event is a duplicate event. Hence, the original event in the dedup table gets updated with the new event's description, severity, and status, and the count field of the older event gets increased by 1.

This duplicate event now gets stored in the archive for later analysis. If there is no matching event found in the dedup table, then the new incoming event represents a new issue and gets stored as a new entry in the dedup table.

After processing all events in the alert bank, let's find out how much noise (duplicate events) was filtered by the deduplication function. We will compare the events present in the archive table and dedup table and then use the Python library matplotlib to plot the pie chart for this analysis.

```
df_dedup = pd.read_sql("select * from dedup" , Database().
getConn())
df_archive = pd.read_sql("select * from archive", Database().
getConn())
fig, ax = plt.subplots()
source = ["Actual-Issues", "Duplicate Event - Noise"]
colors = ['green','yellow']
values = [len(df_dedup), len(df_archive)]
ax.pie(values, labels=source, colors=colors, autopct='%.1f%%',
shadow=True)
plt.show()
```

Based on the pie chart in Figure 6-8, the deduplication function has filtered out 70.6 percent of duplicated events, leaving only 29.4 percent of actual events for IT operations team, which is a considerable reduction in event volume to be processes while retaining the useful information coming via new duplicated events.

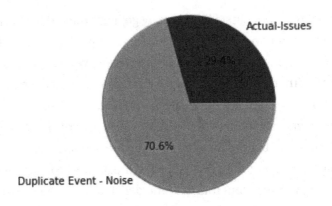

***Figure 6-8.*** *Deduplication function noise removal*

Let's observe what are the top noise maker CIs in data, as shown in Figure 6-6.

```
df_dedup = df_dedup.sort_values(by="Count", ascending=False)
df_dedup[:10].plot(kind="bar", by="Count", x="Host")
plt.title("Top 10 host with most de-duplications", y=1.02);
```

Figure 6-9 shows the top 10 CIs that generate maximum events in the environment that need to be analyzed by problem management and capacity management teams.

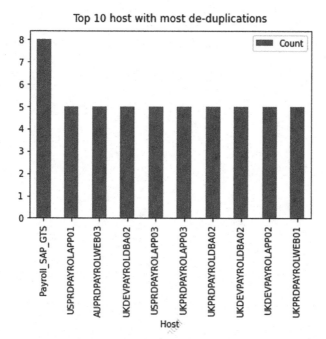

**Figure 6-9.** *Top 10 noisy host*

This completes our first use case of deduplication in AIOps.

# Summary

In this chapter, we covered how to set up the environment using Anaconda for running the use cases. We covered our first use case, which was deduplication. This use case does not use any machine learning but relies on rule-based correlation to deduplicate events. In the next chapter, we will cover another important use case in AIOps: automated baselining.

# CHAPTER 7

# AIOps Use Case: Automated Baselining

We are continuing to discuss specific use cases of AIOps, and in this chapter we explain and implement automated baselining, which is one of the most important and frequently used features in AIOps.

## Automated Baselining Overview

In traditional monitoring tools, a baseline is static and rule-based and is set at predetermined levels. An example of a baseline could be CPU utilization, which is set at, say, 95 percent. An event gets triggered when the CPU utilization of a server goes above this threshold. Now the challenge with this approach is that baselines or thresholds cannot be static. Consider a scenario where an application runs a CPU-intensive data processing job at 2 a.m. local time and causes 99 percent CPU utilization until the job ends, say, at 2:30 a.m. This is the usual behavior of the application, but because of the static baseline threshold, an event gets generated. But this alarm doesn't need any intervention as it will automatically close once the job get completed and the CPU utilization returns to normal. There are many such false positive events in operations due to this static baselining of thresholds.

© Navin Sabharwal and Gaurav Bhardwaj 2022
N. Sabharwal and G. Bhardwaj, *Hands-on AIOps*,
https://doi.org/10.1007/978-1-4842-8267-0_7

Automated baselining helps in such scenarios because it considers the dynamic behavior of systems. It analyzes historical performance data to detect the real performance issues that need attention and eliminates the noise by adjusting the baseline threshold. In the given scenario, the system would automatically increase the threshold and not generate the event until after 2:30 a.m. However, if the same machine exhibits a CPU spike at, say, 9 a.m., it would generate an event.

It is important to note that the dynamic baseline of a threshold will work a bit differently in a microservices architecture where the underlying infrastructure will scale up based on increased utilization or load. These aspects of application architecture and infrastructure utilization need to be considered for dynamic baselining.

Noise in operations causes the following inefficiencies:

- Increases volume of events for operations to manage, which in turn increases time and effort of operation teams.

- Clutters the operator console with false positive events, which leads to missing important qualified and actionable events.

- Causes failures in automated diagnosis and remediation. Since the event itself is false, the automated resolution would be triggered unnecessarily.

Automated baselining reduces noise and related inefficiencies. Automated baselining can be achieved by leveraging supervised machine learning techniques that learn from past data and predict dynamic thresholds based on daily, weekly, monthly, and yearly seasonality of the data. From an AIOps perspective, there are three prominent algorithms that are core to implementing automated baselining. We will explore these approaches in this chapter starting with regression algorithms.

# Regression

These classes of algorithms are used to determine the relationship between multiple variables and predict the value of a target variable. For example, based on the historical analysis of marketing spend, predict the next month's sales revenue. Primarily, linear regression belongs to the statistics domain but is extensively used by machine learning for predictive modeling and is widely useful in multiple domains such as predicting stock prices, sales to price ratio of product, housing prices, etc. There are multiple variants of the regression algorithm, but the linear regression algorithm is one of the most-used regression algorithms in machine learning. Figure 7-1 lists some of the important linear regression algorithms available, and we will be discussing linear regression in the next section.

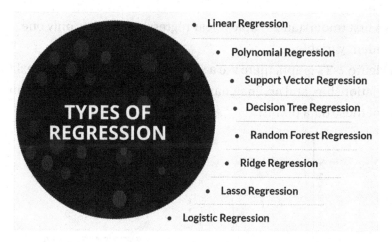

*Figure 7-1.* *Regression algorithms*

# Linear Regression

This algorithm is applicable where the dataset shows a linear relationship between input and output variables and the output variable has continuous values, such as percentage utilization of CPU, etc.

In linear regression, input variables, referred to as *independent variables* or *features*, from the dataset get chosen to predict the values of the output variable, referred to as the *dependent variable* or *target value*. Mathematically, the linear regression model establishes a relationship between the dependent variable (Y) and one or more independent variables (Xi) using a best-fit straight line (called the *regression line*) and is represented by the equation $Y = a + bX_i + e$, where a is the intercept, b is the slope of the line, and e is the error in prediction from the actual value. When there are multiple independent variables, it's called *multiple linear regression*.

Let's first understand simple linear regression that has only one independent variable.

In Figure 7-2's graph, there are a set of actual data points that exhibit a linear relationship, and an imaginary straight line runs somewhere in the middle of these data points.

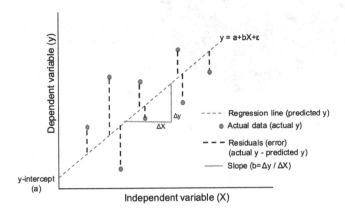

*Figure 7-2.* *Simple linear regression*

This imaginary straight line is the regression line, which consists of predicted data points. Now the distance between each one of the actual data points, and the line is called an *error*. Multiple lines can be drawn that pass through these data points, so the algorithm finds the best line that minimizes this error by calculating distances with all points. Whichever lines gives the smallest error becomes the regression line and provides the best prediction. From an AIOps perspective, linear regression provides a predictive capability for correlation, but it does not show causation.

Now let's understand the implementation of linear regression to forecast the database response time (in seconds) based on the memory utilization (MB) and CPU utilization (percent) of the server. Here we have two independent variables, namely, CPU and memory, so we will be using multiple linear regression models to predict the database response time based on the specific utilization value of CPU and memory.

You can download the code from `https://github.com/dryice-devops/AIOps/blob/main/Ch-7_AutomatedBaselinig-Regression.ipynb`.

Let's begin by exporting data for the parameters CPU Utilization, Memory Utilization, and DB Response Time from the file `https://github.com/dryice-devops/AIOps/blob/main/data.xlsx`, which is shown in Table 7-1.

*Table 7-1.* *CPU and Memory Utilization Mapping with DB Response Time*

| DBResponseTimeSeconds | CPUUsagePercent | MemoryUtilizationMB |
|---|---|---|
| 2.19 | 6.75 | 1215 |
| 2.44 | 6.88 | 1030 |
| 2.56 | 7.00 | 884 |
| 2.69 | 7.13 | 1203 |
| 2.81 | 7.38 | 1193 |
| 2.94 | 7.63 | 1113 |
| 3.06 | 7.75 | 1091 |
| 3.19 | 7.88 | 909 |
| 3.31 | 8.25 | 1218 |
| 3.44 | 8.50 | 1108 |
| 3.56 | 8.75 | 1193 |
| 4.00 | 8.50 | 1324 |
| 4.13 | 8.63 | 1360 |
| 4.25 | 9.13 | 1430 |
| 4.69 | 9.38 | 1478 |
| 4.81 | 9.63 | 1514 |
| 4.94 | 9.75 | 1470 |
| 5.38 | 9.63 | 1590 |
| 5.50 | 10.00 | 1566 |
| 5.63 | 10.13 | 1595 |
| 5.75 | 10.38 | 1769 |
| 5.88 | 10.50 | 1724 |
| 6.00 | 10.63 | 1645 |
| 6.44 | 10.63 | 1860 |

Along with Pandas and Matplotlib, we will be using the additional Python libraries listed here:

- *NumPy*: This is the most widely used library in Python development to perform mathematical operations on matrices, linear algebra, and Fourier transform.

- *Seaborn*: This is another data visualization library that is based on the Matplotlib library and primarily used for generating informative statistical graphics.

- *Sklearn*: This is one of the most important and widely used libraries for machine learning–related development as it contains various tools for modeling and predictions.

Let's begin writing code by importing the required libraries and functions, as shown here:

```
import pandas as pd
import matplotlib.pyplot as plt
import numpy as np
import seaborn as sns
from sklearn.model_selection import train_test_split
from sklearn.linear_model import LinearRegression
from sklearn.metrics import mean_absolute_error
from sklearn.metrics import r2_score
from sklearn.metrics import mean_squared_error
 percentmatplotlib inline
import matplotlib as mpl
import matplotlib.pyplot as plt
```

Now read the performance data from the input file and observe the top five values in the dataset.

```
df = pd.read_excel("data.xlsx")
df.head()
```

Figure 7-3 shows the sample values for CPU and memory utilization and the corresponding database response times.

Out[2]:

| | DBResponseTime | CPUUsage | MemoryUtilizationMB |
|---|---|---|---|
| 0 | 2.1875 | 6.750 | 1215.00 |
| 1 | 2.4375 | 6.875 | 1030.00 |
| 2 | 2.5625 | 7.000 | 883.75 |
| 3 | 2.6875 | 7.125 | 1202.50 |
| 4 | 2.8125 | 7.375 | 1192.50 |

***Figure 7-3.*** *CPU and memory utilization mapping with DB response time*

In the dataset we have utilization values of the parameters DBResponseTime, CPUUsage, andMemoryUtilizationMB as three different columns. Let's find out how many datapoints are present in the dataset for analysis and prediction.

```
df.shape
```

As shown in Figure 7-4, there are a total of 24 rows in the dataset providing values for each of the three independent and dependent variables.

Out[3]:  (24, 3)

***Figure 7-4.*** *Count of data points in DataSet*

One of the important conditions for the applicability of linear regression algorithms is that there must exist a linear relationship between the input and output data points. Let's first find out if the data satisfies this condition by plotting a graph between DB Response Time and CPU Usage.

```
sns.jointplot(x=df['DBResponseTime'], y=df['CPUUsage'], \
              data=df, kind='reg')
```

As per the graph in Figure 7-5, there exists a linear relationship between CPU Utilization and DB Response Time.

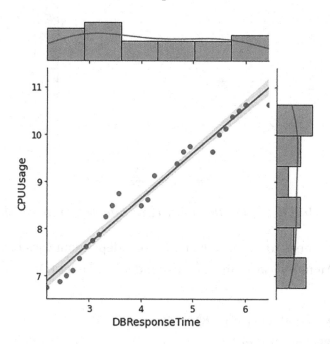

***Figure 7-5.*** *Impact on DB ResponseTime due to CPU utilization*

Similarly, let's validate the linear relationship between DB Response Time and Memory Usage.

```
sns.jointplot(x=df['DBResponseTime'],
y=df['MemoryUtilizationMB'], \
              data=df, kind='reg')
```

Figure 7-6 also shows that a linear relationship exists between DB Response Time and Memory Usage, which satisfies the condition of applying the linear regression algorithm.

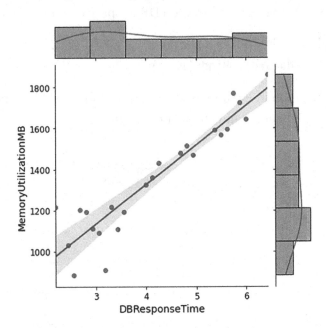

***Figure 7-6.*** *Impact of DB ResponseTime due to memory utilization*

Next, you need to extract data points of independent variables (CPU and Memory usage) in X and dependent variables (DB Response Time) in Y.

```
X = df[['MemoryUtilizationMB','CPUUsage']]
Y = df['DBResponseTime']
```

As discussed in supervised learning, the entire labeled dataset needs to be split into training data for learning purposes and testing data for validating the accuracy or quality of learning done by the model. Now this data needs to be split into testing and training data. In our example, we are considering the ratio of 80 percent to 20 percent for splitting the dataset between testing and training data.

```
x_train, x_test, y_train, y_test = train_test_split(X, Y, \
                    test_size = 0.2, random_state = 42)
```

At this stage, we will invoke LinearRegression from the Sklearn library and apply (called the *fit method*) training data into this linear regression model to create the equation and learn from it.

```
LR = LinearRegression()
# fitting the training data
LR.fit(x_train,y_train)
```

This model can now be used to predict values on the test data.

```
y_prediction =  LR.predict(x_test)
y_prediction
```

Based on the CPU and Memory Utilization values present in the test data, our model has forecasted DB Response Time Values, as shown in the output in Figure 7-7.

```
Out[8]:  array([3.60063185, 5.02933512, 2.49967556, 5.33846418, 3.92287079])
```

***Figure 7-7.*** *Forecasted database response time on test dataset*

Let's compare the predicted values from the ML model with the actual observed values on the test dataset.

```
indices = np.arange(len(y_prediction))
width = 0.20
# Plotting
plt.bar(indices, y_prediction, width=width)
# Offsetting by width to shift the bars to the right
plt.bar(indices + width, y_test, width=width)
plt.xticks(ticks=indices)
plt.ylabel("DB Response Time")
plt.xlabel("Test Dataset ")
```

```
plt.title("Predicted vs. Actual")
valuesType=['Predicted','Actual']
plt.legend(valuesType,loc=2)
plt.show()
```

As observed in the output in Figure 7-8, predicted values are very close to the actual observed value. This indicates that the model is performing well.

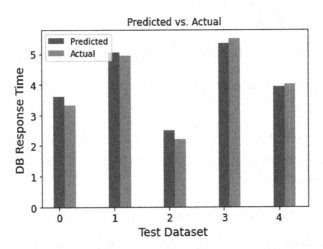

***Figure 7-8.*** *Accuracy analysis of DB Response Time prediction on test data*

Let's evaluate the performance of the model by calculating its R2 score.

```
score=r2_score(y_test,y_prediction)
print("r2 score is ", score)
print('Root Mean Squared Error is =',np.sqrt(mean_squared_
error(y_test,\ y_prediction)))

r2 score is  0.9679128701220477
Root Mean Squared Error is = 0.2102141626503609
```

As per the output, the r2 score value is 0.9679, which means that the model is predicting the database response time with around 97 percent accuracy for a given scenario of CPU and memory utilization, which is quite good. Based on the complexity and requirements, this model can be further expanded to include other independent variables such as the number of transactions, user connections, available disk space, etc., which makes this model production usable to perform capacity planning, upgrades, or migration-related tasks to improve DB response time, which will eventually improve the application performance and user experience.

There are a few limitations with linear regression model where it performs correlation between the dependent variable with one or more independent variables to perform predictions. The first big challenge is that correlations between variables are independent of any time series associated with them. Linear regression cannot predict the value of the dependent variable at a "specific time" with given values of independent variables. Second, prediction is entirely based on the independent variables without considering values of the dependent variable that were already predicted in the recent past. To overcome such challenges, you can use the time-series model, which we will discuss in the next section.

---

**Note**    Technically, a time series can also be considered as an independent variable in linear regressions, but that's not correct. Rather, time-series models should be used instead of linear regression models in such scenarios.

---

# Time-Series Models

To model scenarios that are based on time series and forecast what values that data is going to take into the future at a specified time interval, you can leverage time-series modeling. Let's understand a few important terms required in time-series modeling.

# Time-Series Data

Any data that is collected at regular time intervals is called *time-series data*. Data collection frequency can be anything like hourly, daily, weekly, etc., as long as it is happening at regular intervals. The following are common examples of time-series data:

- *Stock price data*: The maximum and minimum prices of stock are usually maintained at a daily level. It is then further analyzed on a monthly or yearly time frame to make business decisions.

- *Calls in call center*: The call volume at every hour is maintained. It is then analyzed to determine the peak hour load on a monthly or yearly time frame from a resourcing perspective.

- *Sales data*: Sales data is important for any business, and usually monthly sales figures are maintained to forecast future sales or profitability of the organization.

- *Website hits*: Website hit volume is usually maintained every five to ten minutes to determine the load on the application, detect any potential security threats, or scale the infrastructure to meet business requirements.

Every time-series data will have a date or time column along with one or more data columns. If there is only one data column, then it's called a *univariate time series*, and if there are multiple data columns, then it's called a *multivariate time series*. Figure 7-9 is the example of a univariate time series with the daily average CPU utilization of a server.

| TimeStamp | CPUUtilization(Y) |
|---|---|
| 7/1/20 2:00 PM | 8.33 |
| 7/2/20 2:00 PM | 8.82 |
| 7/3/20 2:00 PM | 8.36 |
| 7/4/20 2:00 PM | 11.06 |
| 7/5/20 2:00 PM | 12.52 |
| 7/6/20 2:00 PM | 7.92 |
| 7/7/20 2:00 PM | 8.56 |

***Figure 7-9.*** *CPU utilization time series*

For the purpose of analysis, a univariate time series may need to be converted into multivariate by adding some additional supporting columns. With the help of TimeStamp, we have added another column, Day of the week, and from Figure 7-10 we can observe that there is an unusual spike in CPU utilization on weekends as compared to weekdays.

| TimeStamp | CPUUtilization(Y) | Day of the week |
|---|---|---|
| 7/1/20 2:00 PM | 8.33 | Wednesday |
| 7/2/20 2:00 PM | 8.82 | Thursday |
| 7/3/20 2:00 PM | 8.36 | Friday |
| 7/4/20 2:00 PM | 11.06 | Saturday |
| 7/5/20 2:00 PM | 12.52 | Sunday |
| 7/6/20 2:00 PM | 7.92 | Monday |
| 7/7/20 2:00 PM | 8.56 | Tuesday |

***Figure 7-10.*** *TimeSeries analysis of CPU utilization*

# Stationary Time Series

One of the most important statistical properties of a time series is *stationarity*, where the mean and the variance of the time series do not change over time, as shown in Figure 7-11. In simple terms, a stationary time series can have different values with different timestamps, but the underlying logic or method to generate those values remains constant and does not change.

### Stationary Time Series

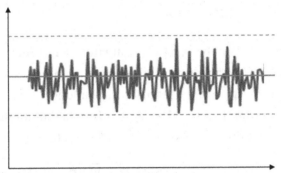

***Figure 7-11.*** *Sample stationary time series*

# Lag Variable

A lag variable is a dependent variable that is lagged in over time. Lag-1 ($Y_{t-1}$) represents the value of dependent variable $Y_t$ at previous one unit of time, Lag 2 ($Y_{t-2}$) will have value of $Y_t$ at previous two units of time, and so on.

Consider the earlier example of time series on CPU utilization at specific timestamps. In this example, Lag-1 ($Y_{t-1}$) at 7/3/20 2:00 PM will be the value of $Y_t$ at 7/2/20 2:00 PM, which is 8.82, and Lag 2 ($Y_{t-2}$) will have value of $Y_t$ at 7/1/20 2:00 PM, which is 8.33, as shown in Figure 7-12.

| Time Stamp | CPU Utilization ($Y_t$) | Lag-1 ($Y_{t-1}$) | Lag-2 ($Y_{t-2}$) |
|---|---|---|---|
| 7/1/20 2:00 PM | 8.33 | - | - |
| 7/2/20 2:00 PM | 8.82 | 8.33 | - |
| 7/3/20 2:00 PM | 8.36 | 8.82 | 8.33 |
| 7/4/20 2:00 PM | 11.06 | 8.36 | 8.82 |
| 7/5/20 2:00 PM | 12.52 | 11.06 | 8.36 |
| 7/6/20 2:00 PM | 7.92 | 12.52 | 11.06 |
| 7/7/20 2:00 PM | 8.56 | 7.92 | 12.52 |

***Figure 7-12.*** *Lag variable and its values*

Mathematically if there exists a relationship or correlation between $Y_t$ and its lagged values $Y_{t-1}$, $Y_{t-2}$, etc., then a model can be created to detect the pattern and predict future values of the same series using its lagged values. For example, assume that CPU utilization increases around 100 percent on every Saturday; then the value at Lag-7 (Yt-7) can be considered for prediction.

The next step is to determine the strength of the relationship between $Y_t$ and its lags and how many lags have statistically significant relationship with $Y_t$ that we can use in the prediction model. That's where ACF and PACF plots help, and we will be discussing these next.

# ACF and PACF

Both ACF and PACF are statistical methods used by researchers to understand the temporal dynamics of an individual time series, which basically means detecting correlation over time, often called *autocorrelation*. A detailed discussion on ACF and PACF is beyond scope of this book, so we will be limiting the discussion to the basics of ACF and PACF to understand how to use them in the AIOps model development.

ACF is a statistical term that represents the autocorrelation function and is used to measure the correlation of time-series values $Y_t$ with its own lags $Y_{t-1}$, $Y_{t-2}$, $Y_{t-3}$, etc. This correlation of various lags can be visualized using an ACF plot on a scale of 1 to -1, where 1 represents a perfect positive relationship and -1 represents a perfect negative relationship.

As shown in Figure 7-13, the y-axis represents the correlation value, and the x-axis represents the lags. The first value of the series will always be 100 percent correlated with itself, and that's why in the ACF plot the Lag 0 correlation value is going until 1. The ACF plot is useful in detecting the pattern in the correlated data.

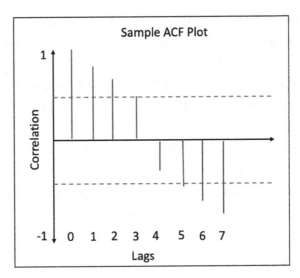

***Figure 7-13.***  *Sample ACF plot*

There is a red line shown on both sides in the ACF plot called the significance line, which is creating a range or band referred to as the *error band*. The lags that cross this line have a statistically significant relationship with variable Y. In a sample ACF plot, three lags are crossing this line, so up to Lag 3, there is a statistically significant relationship, and anything within this band is not statistically significant.

PACF is another statistical term that represents the partial autocorrelation function, which is similar to an ACF plot except that it considers the strength of the relationship between the variable and lag at a specific time after removing the effects of all intermediate lags.

For example, in the sample ACF plot in Figure 7-14, it is visible that there is a significant correlation up to three lags, but the PACF plot tells how much the strength of correlation of $Y_{t-3}$ alone with $Y_t$ is after removing any impact of $Y_{t-1}$ and $Y_{t-2}$. PACF can be considered as more of a filtered value of ACF.

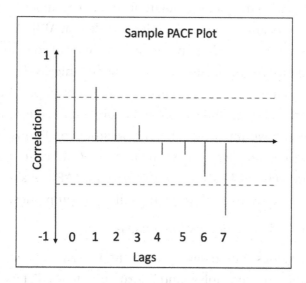

***Figure 7-14.*** *Sample PACF plot*

Now we are all set to implement two most commonly used time-series models, ARIMA and SARIMA, for noise reduction by predicting the appropriate monitoring threshold of parameters. This is one of the most common use cases in an AIOps implementation.

# ARIMA

ARIMA is one of the foundational time-series forecasting techniques; it is popular and has been extensively used for quite a while now. This concept was first introduced by George Box and Gwilym Jenkins in 1976 in a book titled *Time Series Analysis: Forecasting and Control.* They defined a systematic method called the Box: Jenkins Analysis to identify, fit, and use an autoregressive (AR) technique integrated (I) with a moving average (MA) technique resulting in an ARIMA time-series model.

ARIMA leverages the pattern and information contained in a time series to predict its future values. So, is it possible that ARIMA can forecast with any time series? How about predicting the price of your favorite stocks using ARIMA considering its historical prices? Technically, forecasting can be done, but it might not be accurate at all. For example, the price of a stock is controlled by multiple external factors such as the company's quarterly results, government regulations, weather conditions, competitors price, etc. Without considering all such factors, ARIMA cannot provide meaningful predictions. It is important to note that ARIMA should be used only for those time series that exhibit the following properties:

- It should be stationary, or nonseasonal.

- Past values should have a sufficient pattern (not just a random white noise) and information to predict its future values.

- It should be of medium to long length with at least 50 observations.

# Model Development

The ARIMA model has three components.

## Differencing (d)

As discussed earlier, the ARIMA model needs the series to be stationary, but then the challenges are as follows:

- How to attain stationarity in a series

- How to validate if the series becomes stationary or not

There are two approaches to change nonstationary series into stationary.

- Perform an appropriate transformation such as a log transformation or square root transformation to change a nonstationary series into a stationary series that is relatively complex.

- A more reliable and widely used approach is to perform differencing on a series to convert it into stationary. In this approach, the difference between consecutive values in a series gets calculated, and then the resultant series is validated whether or not it is stationary.

One of the most used statistical tests to determine whether a time series is stationary or not is the augmented Dickey Fuller test (ADF test), which is one type of unit root test that determines how strongly a time series is defined by a trend. In simple terms, the ADF test validates a null hypothesis (H0), which claims that a time series has some time-dependent structure and hence it is nonstationary. If this null hypothesis gets rejected, then it means that time series is stationary. The ADF test calculates the p-value, and if it's less than 0.05 (5 percent), then the null hypothesis gets rejected. Interested readers can explore more about internal functioning and mathematics behind unit root tests and ADF tests online.

We need to continue performing differencing and validating the results via an ADF test until we get a near-stationary series with a defined mean. The number of time differencing performed to make time series stationary is represented by model variable d.

It is important to note that series should not get over-differenced in which case we get a stationary series, but the model parameters get affected, and the prediction's accuracy gets impacted. One way to check for over-differencing is that Lag 1 of autocorrelation should not get too negative.

## Autoregression or AR (p)

This component, known as *autoregression*, shows that the dependent variable $Y_t$ can be calculated as a function of its own lags $Y_{t-1}$, $Y_{t-2}$, and so on.

$$Y_t = \alpha + \beta_1 Y_{t-1} + \beta_2 Y_{t-2} + \ldots + \beta_p Y_{t-p} + \epsilon_1$$

Here,

$\alpha$ is the constant term (intercept).

$\beta$ is the coefficient of lag with a subscript indicating its position.

$\epsilon$ is the error or white noise.

We already discussed that the PACF plot shows the partial autocorrelation between the series and its lag after excluding the effect from the intermediate lags. We can use the PACF plot to determine the last lag p, which cuts off the significant line and can be considered as a good estimate value of AR. This value is represented by the model variable p as an order of the AR model.

Based on the sample PACF plot in Figure 7-15, we can observe that lag 1 seems to be the last lag crossing a significant line. So, we can set p as 1 and represent it as AR(1).

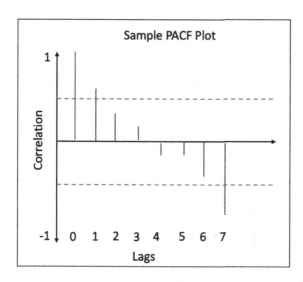

*Figure 7-15.  PACF plot*

# Moving Average or MA (q)

This component, known as *moving average*, shows that the dependent variable $Y_t$ can be calculated as a function of the weighted average of the forecast errors in previous lags.

$$Y_t = \alpha + \epsilon_t + \phi_1\epsilon_{t-1} + \phi_2\epsilon_{t-2} + .. + \phi_q\epsilon_{t-q}$$

α is usually the mean of the series.

ε is the error from the autoregressive models at respective lags.

MA can detect trends and patterns in time-series data. You can use the ACF plots to determine the value of model variable q, which represents how many moving averages are required to remove any autocorrelation in a stationary series. From the ACF plot in Figure 7-16, we can observe that there is cut-off after lags 2 and 6. So, we can try ACF with q =2 , MA (2) or q=6, MA(6).

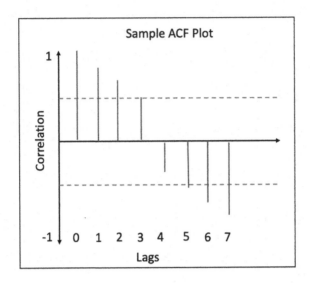

*Figure 7-16.  ACF plot*

Both AR and MA equations needs to be integrated (I or d) to create the ARIMA model, which is represented as ARIMA (p,d,q).

ARIMA acts as a foundation for various other autoregressive time-series models as follows:

- *SARIMA*: If there is a repetitive pattern at a regular time interval, then it's referred to as *seasonality*, and the SARIMA model is used where S stands for seasonality. We will explore this more in the next section.

- *ARIMAX*: If the series depends on external factors, then the ARIMAX model is used, where X stands for external factors.

- *SARIMAX*: If along with seasonality there is a strong influence of external factors, then the SARIMAX model is used.

One of the big limitations of ARIMA is the inability to consider the seasonality in time-series data, which is common in real-world scenarios like when there is exceptionally high sales figures during the holiday period of Thanksgiving and Christmas every year, high expenditure during the first week of every month, high footprints during the weekend in shopping malls, and so on. Such seasonality gets considered in another algorithm, called SARIMA, which is a variant of ARIMA model.

# SARIMA

A cyclical pattern at a regular time interval is referred as *seasonality* and should be considered in a prediction model to improve the accuracy of prediction. That's how the ARIMA model is modified to develop SARIMA, which considers seasonality in time-series data and supports univariate time-series data with a seasonal component.

Along with ARIMA variables, SARIMA needs four new components.

- **p**: Autoregression (AR) order of time series.

- **d**: Difference (I) order of time series.

- **q**: Moving average (MA) order of time series.

- **m (or lag)**: Seasonal factor, number of time steps to be considered for a single seasonal period. For example, if there is yearly seasonality, then m = 12; if it's quarterly, then m=4; if it's weekly, then m=52.

- **P**: Seasonal autoregressive order.

- **D**: Seasonal difference order.

- **Q**: Seasonal moving average order.

The final notation for an SARIMA model is specified as SARIMA (p,d,q) (P,D,Q)m.

# Implementation of ARIMA and SARIMA

Let's perform a sample implementation of the ARIMA and SARIMA models on CPU utilization data to predict its thresholds.

Begin by importing the required Python libraries.

```
# Library for Data Manipulation
import pandas as pd
import numpy as np
```

```
# Library for Data Visualization
import matplotlib.pyplot as plt
```

```
# Library for Time Series Models
from statsmodels.tsa.stattools import adfuller
from statsmodels.graphics.tsaplots import plot_acf, plot_pacf
from statsmodels.tsa.arima.model import ARIMA
from statsmodels.tsa.stattools import acf
from statsmodels.tsa.seasonal import seasonal_decompose
```

```
# Library for Machine Learning
from sklearn.metrics import mean_squared_log_error
```

```
# Library for Track time
from datetime import datetime
```

Read data on CPU utilization from a CSV in a Pandas data frame and check information about the data types, column names, and count of values present in imported data.

```
cpu_util_raw_data= pd.read_csv("cpu_utilization-data.csv")
cpu_util_raw_data.info()
```

As shown in Figure 7-17, there are two columns, date_time and cpu_utilization, and a total of 396 data points (first row is the header) are available for analysis.

```
<class 'pandas.core.frame.DataFrame'>
RangeIndex: 397 entries, 0 to 396
Data columns (total 2 columns):
 #    Column               Non-Null Count   Dtype
---   ------               --------------   -----
 0    date_time            397 non-null     object
 1    cpu_utilization      397 non-null     float64
dtypes: float64(1), object(1)
memory usage: 6.3+ KB
```

***Figure 7-17.*** *Overview of data in data file*

Set the data type of column date_time to DateTime, set it as an index to sort the data points in a data frame, and plot the CPU utilization on a time scale, as shown in Figure 7-18.

```
total_records = cpu_util_raw_data.count()['cpu_utilization']
cpu_util_raw_data['date_time'] = pd.to_datetime(cpu_util_raw_
data['date_time'])
cpu_util_raw_data.set_index('date_time',inplace=True)
cpu_util_raw_data = cpu_util_raw_data.sort_
values(by="date_time")
fig = plt.figure(figsize =(10, 5))
cpu_util_raw_data['cpu_utilization'].plot()
```

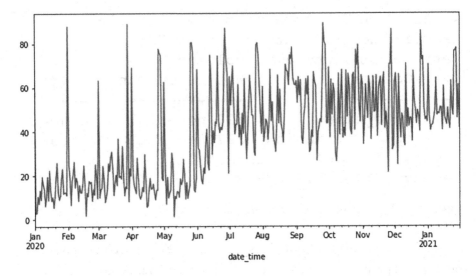

***Figure 7-18.*** *CPU utilization on time scale*

It is important to check for the outliers in the data using a box plot. The presence of outliers impacts the quality of predictions. If there are a high number of outliers in data, then they need to be either processed or removed from the dataset based on the domain requirements.

```
#Global variables
NOISE = False
MAPE = True
TEST_DATA_SIZE = 0.2 #in percent
SEASONAL = False
def getOutliers(data, col):
    Q3 = data[col].quantile(0.75)
    Q1 = data[col].quantile(0.25)
    IQR = Q3 - Q1
    lower_limit = Q1 - 1.5 * IQR
    upper_limit = Q3 + 1.5 * IQR
    return lower_limit, upper_limit
```

```
lower, upper = getOutliers(cpu_util_raw_data, 'cpu_
utilization')
outliers_count = cpu_util_raw_data[(cpu_util_raw_
data['cpu_utilization']<lower) | (cpu_util_raw_data['cpu_
utilization']>upper)].count()['cpu_utilization']
outlier_percentage = ((outliers_count / total_records) * 100)

if outlier_percentage > 20:
    NOISE = True

print(f"Outlier Percentage in data: {outlier_percentage}%")

#Render box plot
cpu_util_raw_data.boxplot('cpu_utilization', figsize=(10,10))

zero_values = cpu_util_raw_data[(cpu_util_raw_data['cpu_
utilization']==0)].count()['cpu_utilization']
zero_values_percentage = ((zero_values / total_records) * 100)

if zero_values_percentage > 20:
    MAPE = False

print("Zero Value percentage in data: ", zero_values_
percentage)
```

Based on Figure 7-19, there are no outliers and no "zero" values, so let's continue with the given dataset.

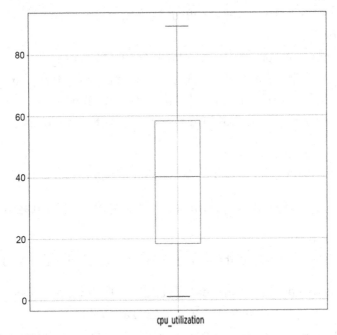

```
Zero Value percentage in data:   0.0
```

***Figure 7-19.*** *Box plot analysis for outliers*

Let's check for the stationarity of a time series using an ADF test. Based on an ADF test, if the p-value comes out to more than 0.05, it means the time series is nonstationary and needs to be differenced using the diff() function and again checked with an ADF test for stationarity.

```
diff_count = 0
differencing_order = {
    1: lambda x: x['cpu_utilization'].diff(),
    2: lambda x: x['cpu_utilization'].diff().diff(),
    3: lambda x: x['cpu_utilization'].diff().diff().diff(),
    4: lambda x: x['cpu_utilization'].diff().diff().
    diff().diff(),
    5: lambda x: x['cpu_utilization'].diff().diff().diff().
    diff().diff()
```

```
}
while True:
    if diff_count == 0:
        adftestresult = adfuller(cpu_util_raw_data['cpu_
        utilization'].dropna())
    else:                                              .
        adftestresult = adfuller(differencing_order[diff_count]
        (cpu_util_raw_data).dropna())

    print('#' * 60)
    print('ADF Statistic: %f' % adftestresult[0])
    print('p-value: %f' % adftestresult[1])
    print(f'ADF Test Result: The time series is {"non-" if
    adftestresult[1] >= 0.05 else ""}stationary')
    print('#' * 60)

    if adftestresult[1] < 0.05 or diff_count >=
    len(differencing_order):
        break
    diff_count += 1

print("Differencing order to make data stationary: ",diff_count)
```

```
############################################################
ADF Statistic: -2.533679
p-value: 0.107493
ADF Test Result: The time series is non-stationary
############################################################
############################################################
ADF Statistic: -15.396157
p-value: 0.000000
ADF Test Result: The time series is stationary
############################################################
Differencing order to make data stationary:  1
```

The ADF result on differenced time series shows that the p-value is 0, which confirms that data series is now stationary. This can be visualized by plotting both the original series and a differenced series on a time scale graph, as shown in Figure 7-20.

```
fig, ax = plt.subplots(figsize=(10,8), dpi=100)
# Differencing
ax.plot(cpu_util_raw_data.cpu_utilization[:], label='Original
Series')
ax.plot(cpu_util_raw_data.cpu_utilization.diff(1), label='1st
Order Differencing')
ax.set_title('1st Order Differencing')
ax.legend(loc='upper left', fontsize=10)
plt.legend(loc='upper left', fontsize=10)
plt.suptitle('CPU Usage - Time Series Dataset', fontsize=16)
plt.show()
```

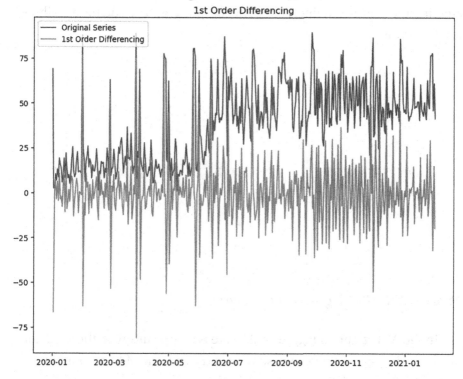

**Figure 7-20.** *Time-series stationarity check*

As we have done differencing only one time, we can set the value of the differencing (d) component of ARIMA as d=1.

Next let's plot the ACF and PACF graph to get the value of the AR (p) and MA (q) components.

```
plt.rcParams.update({'figure.figsize':(7, 4), 'figure.
dpi':120})
plot_pacf(cpu_util_raw_data.cpu_utilization.diff().dropna());
plot_acf(cpu_util_raw_data.cpu_utilization.diff().dropna());
```

In the PACF graph in Figure 7-21, it can be observed that until lag 6 it is crossing the significant line, so the AR component, p=6, can be initially set for modeling.

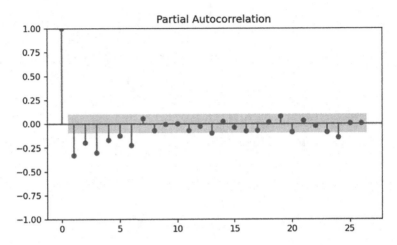

***Figure 7-21.*** *PACF graph of time series*

In the ACF graph in Figure 7-22, there is sharp cut-off on the lag 2 and lag 7 crossing significance line, so we can try with the MA component as q=2 or q =7 for modeling.

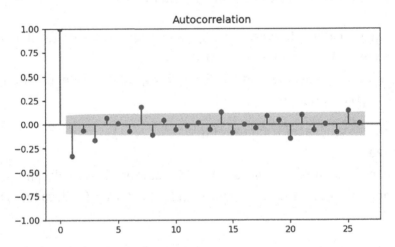

***Figure 7-22.*** *ACF graph of time series*

Next split the data into training and test data to be used for the ARIMA model (6,1,2). Out of 396 data points, let's use the first 300 (~75 percent) data points for training and the rest of the data points for testing purposes and plot them on a time scale, as shown in Figure 7-23.

```
# Plot train and test data

train_data = cpu_util_raw_data['cpu_utilization'][:300]
test_data = cpu_util_raw_data['cpu_utilization'][300:]
fig, ax = plt.subplots(figsize=(10,5), dpi=100)

ax.plot(train_data, label='Train Data Points')
ax.plot(test_data, label='Test Data Points')
ax.set_title('CPU Usage Timeline')
ax.legend(loc='upper left', fontsize=10)
plt.legend(loc='upper left', fontsize=10)
plt.suptitle('Test-Train Data Splits', fontsize=16)
plt.show()
```

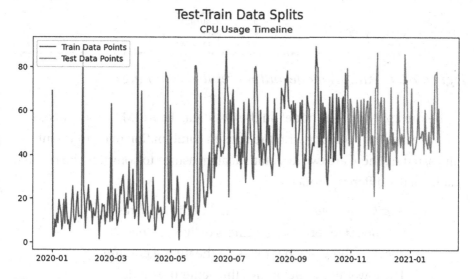

**Figure 7-23.** *Time-series data split in test data and training data*

Let's create the ARIMA (6,1,2) model using the train data.

```
manual_arima_model = ARIMA(train_data, order=(6,1,2))
manual_arima_model_fit = manual_arima_model.fit()
print(manual_arima_model_fit.summary())
```

Figure 7-24 shows the output from the ARIMA model on the train data.

```
                              SARIMAX Results
==============================================================================
Dep. Variable:         cpu_utilization   No. Observations:              300
Model:                  ARIMA(6, 1, 2)   Log Likelihood            -1254.120
Date:                Thu, 07 Apr 2022   AIC                        2526.240
Time:                        19:31:19   BIC                        2559.544
Sample:                    01-01-2020   HQIC                       2539.570
                         - 10-26-2020
Covariance Type:                  opg
==============================================================================
                 coef    std err          z      P>|z|      [0.025      0.975]
------------------------------------------------------------------------------
ar.L1         -0.4883      0.345     -1.415      0.157      -1.165       0.188
ar.L2          0.2784      0.134      2.074      0.038       0.015       0.541
ar.L3         -0.1431      0.081     -1.766      0.077      -0.302       0.016
ar.L4         -0.0464      0.092     -0.504      0.615      -0.227       0.134
ar.L5          0.1239      0.078      1.598      0.110      -0.028       0.276
ar.L6         -0.0047      0.077     -0.061      0.952      -0.156       0.147
ma.L1         -0.1293      0.346     -0.374      0.709      -0.807       0.549
ma.L2         -0.7522      0.328     -2.291      0.022      -1.396      -0.109
sigma2       256.0437     12.450     20.566      0.000     231.642     280.445
===================================================================================
Ljung-Box (L1) (Q):                   0.09   Jarque-Bera (JB):            263.18
Prob(Q):                              0.77   Prob(JB):                      0.00
Heteroskedasticity (H):               0.82   Skew:                          1.24
Prob(H) (two-sided):                  0.33   Kurtosis:                      6.87
===================================================================================
```

***Figure 7-24.*** *ARIMA model output result on training data*

Before applying this model on the test data, the ARIMA model needs to be checked for accuracy with the chosen combination of component values, p,d,q. There are two key statistical measures to compare the relative quality of the different models.

- *Akaike information criteria (AIC)*: This validates the model in terms of goodness of fit of the data by checking how much it relies on the tuning parameters. The lower the value of AIC, the better the model performance. The current ARIMA (6,1,2) model has an AIC value of 2113.

- *Bayesian information criteria (BIC)*: In addition to AIC, the BIC uses the number of samples in the training dataset that are used for fitting. Here also a model with a lower BIC is preferred. The current ARIMA (6,1,2) model has a BIC value of 2150.

Based on PACF and ACF charts, multiple values can be tried to lower the AIC and BIC values. For now let's proceed with ARIMA (6,1,2).

Let's plot the actual values in a training dataset with the model's predicted value to analyze how close the actual and model predicted values are.

```
prediction_manual=manual_arima_model_fit.predict(dynamic=False,
typ='levels')
plt.figure(figsize=(15,5), dpi=100)
fig, ax = plt.subplots(figsize=(10,5), dpi=100)
ax.plot(prediction_manual, label='Prediction')
ax.plot(train_data, label='Training Data')
ax.legend(loc='upper left', fontsize=10)
plt.legend(loc='upper left', fontsize=10)
plt.suptitle('Actual V\S Predicted CPU Utilization on Train
Data', fontsize=16)
plt.show()
```

As observed in Figure 7-25, the model predictions are quite close to actual values, indicating that it gets trained quite well.

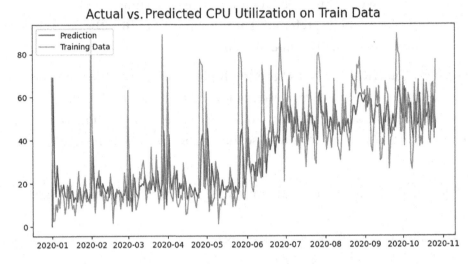

**Figure 7-25.** *Prediction analysis on training data*

Now let's validate the model performance on the test data.

```
# Forecast with 95% confidence interval
forecast = manual_arima_model_fit.get_forecast(97)
manual_arima_fc = forecast.predicted_mean
manual_arima_conf = forecast.conf_int(alpha=0.05)
manual_arima_fc_series = pd.Series(manual_arima_fc, index=test_
data.index)
manual_arima_lower_series = pd.Series(manual_arima_conf["lower
cpu_utilization"], index=manual_arima_conf.index)
manual_arima_upper_series = pd.Series(manual_arima_conf["upper
cpu_utilization"], index=manual_arima_conf.index)
plt.figure(figsize=(15,10), dpi=100)
plt.plot(train_data, label='Training Data')
plt.plot(test_data, label='Actual Data')
```

```
plt.plot(manual_arima_fc_series, label='Forecast Data')
plt.fill_between(manual_arima_lower_series.index, manual_arima_
lower_series, manual_arima_upper_series, color='gray', alpha=.15)
plt.title('Forecast vs Actuals')
plt.legend(loc='upper left', fontsize=8)
plt.show()
```

Figure 7-26 shows that the ARIMA (6,1,2) model performance on the test dataset is not good and its predictions are deteriorating over time.

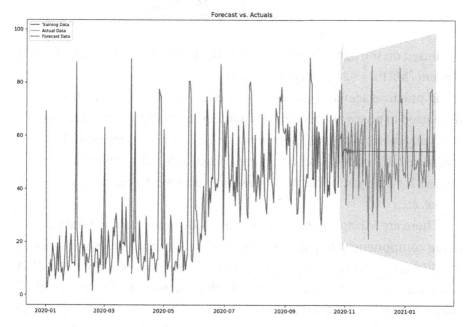

***Figure 7-26.*** *Prediction analysis on test data*

Let's quantify the model performance with the metrics defined earlier.

```
def forecast_accuracy(forecast, actual):
    mape = np.mean(np.abs(forecast - actual)/
    np.abs(actual)*100)
    mae = np.mean(np.abs(forecast - actual))
```

```
rmse = np.mean((forecast - actual)**2)**.5
rmsle = np.sqrt(mean_squared_log_error(actual, forecast))
return({'MAPE : ':mape, 'MAE : ': mae,
        'RMSE : ':rmse, 'RMSLE : ':rmsle
    })
```

```
forecast_accuracy(manual_arima_fc, test_data)
```

```
Out[16]: {'MAPE : ': 24.071440955749733,
          'MAE : ': 11.286266474840719,
          'RMSE : ': 13.201605001517883,
          'RMSLE : ': 0.2593143147566868}
```

Based on the performance metrics, the model predictions are about 76 percent (MAPE = ~24 percent) accurate.

In practical scenarios, multiple values of ARIMA components need to be used to arrive at a model with maximum accuracy in predictions. Manually performing this task requires a lot of time and expertise in statistical techniques. Also, it becomes practically impossible to create models for a large number of entities such as servers, parameters, stocks, etc.

There are ML approaches called Auto ARIMA to learn optimal values of the components p,d,q. In Python there is a library called pdarima that provides the function auto_arima to try various values of ARIMA/SARIMA components in model creation and search for most optimal model with the lowest AIC and BIC values.

Let's try auto ARIMA to find the most optimal component values for the ARIMA model from the given training dataset. In the auto ARIMA model, we currently specify seasonal to be False to explore more about ARIMA. This value will be true for the SARIMA model to consider seasonality. We will also provide the maximum values of p and q that auto ARIMA can explore to obtain the best model. Let's proceed with max_p and max_q as 15 to limit the complexity of the model.

```
# Fit auto_arima on train set
auto_arima_model = pm.auto_arima(train_data, start_p = 1,
start_q = 1, max_p = 15, max_q = 15, seasonal = False,
 d = None, trace = True, error_action ='ignore',
 suppress_warnings = True, stepwise = True)
```

```
# To print the summary
auto_arima_model.summary()
```

As per the auto ARIMA result shown in Figure 7-27, the best model is identified as ARIMA (6,1,0) with marginal improvement in the AIC and BIC values as compared to the previous ARIMA (6,1,2) model.

```
Performing stepwise search to minimize aic
 ARIMA(1,1,1)(0,0,0)[0] intercept   : AIC=2521.912, Time=0.17 sec
 ARIMA(0,1,0)(0,0,0)[0] intercept   : AIC=2625.601, Time=0.01 sec
 ARIMA(1,1,0)(0,0,0)[0] intercept   : AIC=2594.071, Time=0.09 sec
 ARIMA(0,1,1)(0,0,0)[0] intercept   : AIC=2544.440, Time=0.05 sec
 ARIMA(0,1,0)(0,0,0)[0]             : AIC=2623.602, Time=0.01 sec
 ARIMA(2,1,1)(0,0,0)[0] intercept   : AIC=2523.890, Time=0.18 sec
 ARIMA(1,1,2)(0,0,0)[0] intercept   : AIC=2523.900, Time=0.18 sec
 ARIMA(0,1,2)(0,0,0)[0] intercept   : AIC=2525.326, Time=0.11 sec
 ARIMA(2,1,0)(0,0,0)[0] intercept   : AIC=2587.411, Time=0.11 sec
 ARIMA(2,1,2)(0,0,0)[0] intercept   : AIC=2524.835, Time=0.24 sec
 ARIMA(1,1,1)(0,0,0)[0]             : AIC=2523.439, Time=0.05 sec

Best model:  ARIMA(1,1,1)(0,0,0)[0] intercept
Total fit time: 1.215 seconds
```

SARIMAX Results

| Dep. Variable: | y | No. Observations: | 300 |
|---|---|---|---|
| Model: | SARIMAX(1, 1, 1) | Log Likelihood | -1256.956 |
| Date: | Thu, 07 Apr 2022 | AIC | 2521.912 |
| Time: | 19:36:39 | BIC | 2536.714 |
| Sample: | 0 | HQIC | 2527.837 |
| | - 300 | | |
| Covariance Type: | opg | | |

| | coef | std err | z | P>\|z\| | [0.025 | 0.975] |
|---|---|---|---|---|---|---|
| intercept | 0.0984 | 0.050 | 1.962 | 0.050 | 7.98e-05 | 0.197 |
| ar.L1 | 0.3191 | 0.045 | 7.041 | 0.000 | 0.230 | 0.408 |
| ma.L1 | -0.9638 | 0.018 | -52.751 | 0.000 | -1.000 | -0.928 |
| sigma2 | 260.6387 | 17.423 | 14.959 | 0.000 | 226.489 | 294.788 |

| | | | |
|---|---|---|---|
| Ljung-Box (L1) (Q): | 0.18 | Jarque-Bera (JB): | 254.60 |
| Prob(Q): | 0.67 | Prob(JB): | 0.00 |
| Heteroskedasticity (H): | 0.81 | Skew: | 1.28 |
| Prob(H) (two-sided): | 0.28 | Kurtosis: | 6.73 |

*Figure 7-27. Prediction analysis on test data*

Next validate the performance of ARIMA (1,1,1) only on test data.

```
auto_arima_predictions = pd.Series(auto_arima_model.
predict(len(test_data)))
actuals = test_data.reset_index(drop = True)
auto_arima_predictions.plot(legend = True,label = "ARIMA
Predictions", xlabel = "Index",ylabel = "CPU Utilization",
 figsize=(10, 7)) actuals.plot(legend = True, label = "Actual");
forecast_accuracy(np.array(auto_arima_predictions), test_data)
```

As observed in Figure 7-28 and as the performance metrics are obtained, there is even further degradation in the accuracy of the ARIMA (1,1,1) model.

```
{'MAPE  :  ': 36.46708478630482,
 'MAE  :  ': 15.801136939985394,
 'RMSE  :  ': 18.25104146679108,
 'RMSLE  :  ': 0.34741799578019483}
```

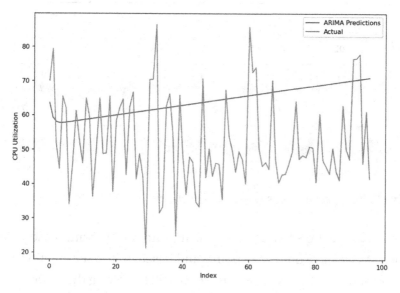

***Figure 7-28.***  *Prediction analysis on test data for ARIMA (1,1,1)*

Clearly, the ARIMA model is not going to work in this scenario, and we need to look for the seasonality in data using the `seasonal_decompose` package.

```
seasonality_check = seasonal_decompose(cpu_util_raw_data
['cpu_utilization'], model='additive',extrapolate_trend='freq')
seasonality_check.plot()

plt.show()
```

Based on the graph in Figure 7-29, we can observe seasonality in the data, which implies we should use the SARIMA model.

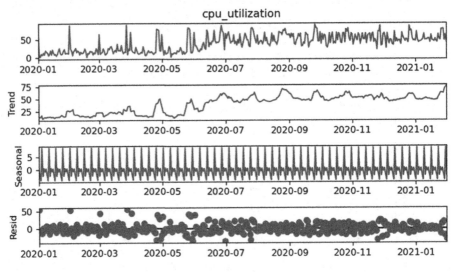

***Figure 7-29.***  *Time series seasonality check*

Let's explore the SARIMA model using auto ARIMA with seasonality enabled. As we have daily data points and based on the seasonality graph, it seems there is weekly seasonality, so we will be setting the value of the SARIMA component as m = 7.

```
# Seasonal - fit stepwise auto-ARIMA
sarima_model = pm.auto_arima(train_data, start_p=1, start_q=1,
```

```
            test='adf',
            max_p=15, max_q=15, m=7,
            start_P=0, seasonal=True,
            d=None, D=1, trace=True,
            error_action='ignore',
            suppress_warnings=True,
            stepwise=True)
sarima_model.summary()
```

As shown in Figure 7-30, auto ARIMA tried various combination and has detected SARIMA (3,0,2)(0,1,1)[7] as a best-fit model with considerable improvement in the AIC values as compared to previous ARIMA models.

```
Performing stepwise search to minimize aic
ARIMA(1,0,1)(0,1,1)[7] intercept   : AIC=inf, Time=0.31 sec
ARIMA(0,0,0)(0,1,0)[7] intercept   : AIC=2641.894, Time=0.01 sec
ARIMA(1,0,0)(1,1,0)[7] intercept   : AIC=2554.842, Time=0.13 sec
ARIMA(0,0,1)(0,1,1)[7] intercept   : AIC=inf, Time=0.23 sec
ARIMA(0,0,0)(0,1,0)[7]             : AIC=2640.437, Time=0.01 sec
ARIMA(1,0,0)(0,1,0)[7] intercept   : AIC=2614.362, Time=0.11 sec
ARIMA(1,0,0)(2,1,0)[7] intercept   : AIC=2535.229, Time=0.34 sec
ARIMA(1,0,0)(2,1,1)[7] intercept   : AIC=inf, Time=0.35 sec
ARIMA(1,0,0)(1,1,1)[7] intercept   : AIC=inf, Time=0.25 sec
ARIMA(0,0,0)(2,1,0)[7] intercept   : AIC=2575.157, Time=0.47 sec
ARIMA(2,0,0)(2,1,0)[7] intercept   : AIC=2537.094, Time=0.43 sec
ARIMA(1,0,1)(2,1,0)[7] intercept   : AIC=2537.142, Time=0.47 sec
ARIMA(0,0,1)(2,1,0)[7] intercept   : AIC=2541.395, Time=0.43 sec
ARIMA(2,0,1)(2,1,0)[7] intercept   : AIC=2538.860, Time=0.81 sec
ARIMA(1,0,0)(2,1,0)[7]             : AIC=2534.163, Time=0.15 sec
ARIMA(1,0,0)(1,1,0)[7]             : AIC=2553.414, Time=0.09 sec
ARIMA(1,0,0)(2,1,1)[7]             : AIC=2500.996, Time=0.45 sec
ARIMA(1,0,0)(1,1,1)[7]             : AIC=2499.056, Time=0.30 sec
ARIMA(1,0,0)(0,1,1)[7]             : AIC=2498.805, Time=0.12 sec
ARIMA(1,0,0)(0,1,0)[7]             : AIC=2612.610, Time=0.04 sec
ARIMA(1,0,0)(0,1,2)[7]             : AIC=2499.133, Time=0.28 sec
ARIMA(1,0,0)(1,1,2)[7]             : AIC=2501.024, Time=0.71 sec
ARIMA(0,0,0)(0,1,1)[7]             : AIC=2545.393, Time=0.07 sec
ARIMA(2,0,0)(0,1,1)[7]             : AIC=2500.761, Time=0.15 sec
ARIMA(1,0,1)(0,1,1)[7]             : AIC=2500.771, Time=0.19 sec
ARIMA(0,0,1)(0,1,1)[7]             : AIC=2508.489, Time=0.11 sec
ARIMA(2,0,1)(0,1,1)[7]             : AIC=2500.645, Time=0.26 sec
ARIMA(1,0,0)(0,1,1)[7] intercept   : AIC=inf, Time=0.33 sec

Best model:  ARIMA(1,0,0)(0,1,1)[7]
Total fit time: 7.613 seconds
```

SARIMAX Results

| Dep. Variable: | y | No. Observations: | 300 |
|---|---|---|---|
| Model: | SARIMAX(1, 0, 0)x(0, 1, [1], 7) | Log Likelihood | -1246.403 |
| Date: | Thu, 07 Apr 2022 | AIC | 2498.805 |
| Time: | 19:43:18 | BIC | 2509.846 |
| Sample: | 0 | HQIC | 2503.227 |
| | - 300 | | |
| Covariance Type: | opg | | |

| | coef | std err | z | P>|z| | [0.025 | 0.975] |
|---|---|---|---|---|---|---|
| ar.L1 | 0.4152 | 0.038 | 11.018 | 0.000 | 0.341 | 0.489 |
| ma.S.L7 | -0.8053 | 0.042 | -19.198 | 0.000 | -0.888 | -0.723 |
| sigma2 | 282.6997 | 14.047 | 20.125 | 0.000 | 255.167 | 310.232 |

| | | | |
|---|---|---|---|
| Ljung-Box (L1) (Q): | 0.42 | Jarque-Bera (JB): | 198.51 |
| Prob(Q): | 0.52 | Prob(JB): | 0.00 |
| Heteroskedasticity (H): | 0.61 | Skew: | 1.22 |
| Prob(H) (two-sided): | 0.02 | Kurtosis: | 6.21 |

*Figure 7-30.* Auto Arima model execution result on train data

Let's validate the results of this new SARIMA model.

```
sarima_predictions = pd.Series(sarima_model.
predict(len(test_data)))
actuals = test_data.reset_index(drop = True)
sarima_predictions.plot(legend = True,label = "SARIMA
Predictions",xlabel = "Index", ylabel = "CPU Utilization",
figsize=(10, 7))
auto_arima_predictions.plot(legend = True,label = "ARIMA
Predictions")
actuals.plot(legend = True, label = "Actual")

forecast_accuracy(np.array(sarima_predictions), test_data)
```

After considering the seasonality, the prediction's accuracy improved to about 62 percent as MAPE reduced to about 38 percent. This improvement in accuracy can also be observed in Figure 7-31 by plotting predictions of the ARIMA and SARIMA models against the actual test data observed.

```
{'MAPE : ': 21.28101642137203,
 'MAE : ': 10.223769472384639,
 'RMSE : ': 13.357000496019426,
 'RMSLE : ': 0.25947913355175073}
```

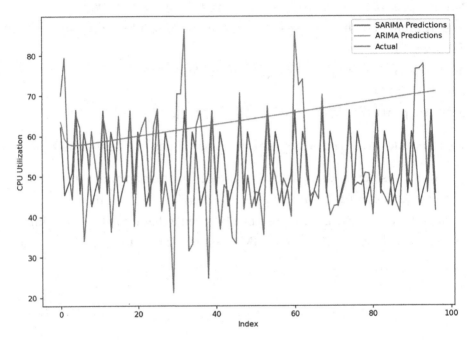

***Figure 7-31.*** *Time-series prediction by ARIMA/SARIMA on test data*

This model can be used to generate future predictions to determine an appropriate baseline dynamically rather than going with a static global baseline. This approach drastically reduces the noise in large environments.

This model can also be used for one of the most common use cases around the autoscaling of infrastructure, especially when there is seasonality in the utilization. The SARIMA (or SARIMAX) model can be used to analyze historical time-series data along with seasonality and any external factor to predict the utilization and accordingly scale up or down infrastructure for a specific duration and provide potential cost savings.

# Automated Baselining in APM and SecOps

Monitoring tools related to SecOps and application monitoring have different characteristics as the majority of time their monitoring parameter values remain very low, almost touching zero, causing imbalanced dataset and hence providing incorrect predictions. In these situations, it makes more sense to compare spikes or unusual high utilization values (called *anomaly*) with predicted values to calculate the MASE ratio and accordingly tune the ML model rather than leveraging the usual low values of parameters. Anomaly values are the ones that do not follow a regular pattern and show sudden spikes (up or down). We will discuss more about these in Chapter 8.

These predicted thresholds for security-related parameters can be applied on policies that monitor the protected segment comprising the IP address range, port, and protocol that define services, VLAN numbers, or MPLS tags. An appropriate policy will dynamically detect attacks on different protected segments and trigger qualified alarms rather than noise.

Implementing a monitoring solution with an automated baselining feature brings in immediate benefits by enabling the operations team to quickly identify outages, exceptions, and cyberattacks rather than wasting time on noise. But organizations often face some operational challenges in adopting dynamic thresholding, which we will be discussing next.

# Challenges with Dynamic Thresholding

Though dynamic thresholding looks promising, there is a challenge in adoption due to fear of missing critical alerts. Especially during the initial phases when the AIOps system has just started the learning process, organizations raise a lot of doubt over the accuracy of prediction. Also, the majority of open source and native monitoring tools don't have this capability. Implementing AIOps-based dynamic thresholding involves the use of multiple algorithms and techniques and requires data for a longer duration to be able to analyze the seasonality and patterns.

# Summary

In this chapter, we covered one of the important use cases in AIOps, which is automated baselining. We covered various types of regression algorithms that can be used for this purpose. We covered hands-on implementation of the use case using multiple algorithms such as linear regression, ARIMA, and SARIMA. In the next chapter, we will cover various anomaly detection algorithms and how they can be used in AIOps.

# CHAPTER 8

# AIOps Use Case: Anomaly Detection

After discussing deduplication and automated baselining in previous chapters, we will now move up the AIOps maturity ladder to discuss anomaly detection, which provides a big leap toward proactiveness. This chapter explains anomaly detection and how it is useful for IT operations.

## Anomaly Detection Overview

Anomaly detection is the process of identifying data points that are unusual or unexpected. Regular events of the CPU, memory, swap memory, disk, etc., are normal to operations, but if there is any "application down" or firewall event, then it represents an unusual scenario. The goal of anomaly detection is to identify such unusual scenarios (what we call *outliers*) within a dataset that differ significantly from other observations. Though the task of detecting an anomaly can be performed by all three types of machine learning algorithms, its implementation is done extensively on unlabeled data by clustering them into various groups. In IT operations, the execution of a service improvement plan is a regular exercise, and operations do not have a specific target to predict. Rather, they need to analyze lots of data and then try to observe similarities and club together them to form different groups to understand anomalies and formulate recommendations. We

© Navin Sabharwal and Gaurav Bhardwaj 2022
N. Sabharwal and G. Bhardwaj, *Hands-on AIOps*,
https://doi.org/10.1007/978-1-4842-8267-0_8

briefly discussed different clustering algorithms available in unsupervised machine learning in Chapter 5 for anomaly detection, and K-means clustering is one of the simplest and most popular unsupervised machine learning algorithms that we will be discovering in this chapter.

## K-Means Algorithms

As discussed in Chapter 5, K-means is a centroid-based clustering algorithm that is extensively used in data mining to detect patterns and group together similar data points into a *cluster*. Mathematically, if there is a dataset {x1, . . . , xn} consisting of N data points, as shown in Figure 8-1, then our goal is to partition the dataset into some number K of clusters.

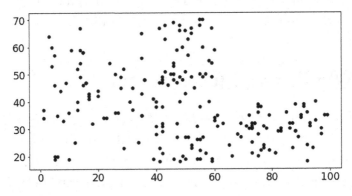

***Figure 8-1.*** *N data points in space*

Here, K represent the number of distinct clusters that can be created around centroids. There are ways to determine the centroids from the dataset, and this section will use the Elbow method in our implementation to determine the appropriate number of clusters. There are three centroids detected in the dataset that are marked as red dots in Figure 8-2.

The K-means algorithm tries to determine the similarity of data points with centroids and accordingly cluster the data points around these centroids, as shown in Figure 8-2.

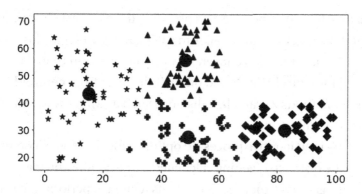

***Figure 8-2.*** *Clustering performed by the K-means algorithm*

The IT operations team performs the task of manually filtering hundreds of events and selecting related events in a group for various tasks such as the following:

- *Root-cause analysis*: Groups events that are related to an issue or outage to find how an issue starts unfolding

- *Performance analysis*: Groups events related to a specific application or service to analyze its performance and related bottlenecks

- *Service improvement*: Groups events that create noise and analyzes what monitoring configurations or baselines should be updated at the source to reduce noise and improve service quality

- *Capacity planning*: Groups events that represent hot spots that need timely intervention and resolution by a capacity planning process

From an AIOps perspective, it is probably the simplest clustering algorithm that can detect anomalies in a wide range of scenarios. In this algorithm, first a centroid is chosen for every cluster, and then data points are grouped into the cluster based on their distance from centroids. There are multiple methods to calculate distance such as Minkowski, Manhattan, Euclidean, Hamming, etc.

Clustering is an iterative process to optimize the position of centroids with which the distance is getting calculated. For example, the majority of organizations receive thousands of events related to performance or availability, but rarely do they receive security alerts related to DDoS/DoS attacks. Not only the frequency but the alarm text message will differ. In this case, you can apply K-means clustering on event messages that group related messages over a time scale, leaving aside security alerts as an anomaly.

Let's consider an implementation scenario where there are multiple alerts in the system and you want to analyze alert messages to segregate frequently occurring events as well as anomalies and then try to determine the root cause of events. For this implementation, you will work with text data; hence, you have to use the subdomain of AI called *natural language processing* (NLP). You will be using NLP to analyze the text by tokenizing it and getting the relevance of tokens and their similarities in different events to create clusters.

To begin, you need to import some libraries. In this implementation, you will be using NLP-related libraries.

- *Natural Language ToolKit (nltk)*: This is a suite of libraries for tokenization, parsing, classification, stemming, tagging, and semantic reasoning of text to process human language data and use it for further analysis in statistical or ML systems.

- *sklearn*: This is a machine learning library that provides the implementation of various algorithms such as K-means, SVM, random forests, gradient boosting, etc. This library allows you to use these algorithms on data rather than making a huge effort to manually implement them and then use them on data.

- *Genism*: This is an important library for topic modeling, *which is a statistical model for text mining and discovering abstract topics that are hidden in a collection* of text. *Topics* can be defined as a repeating pattern of co-occurring terms in a text corpus. For example *switch, interface, router,* and *bandwidth* are terms that will occur together frequently and hence can be grouped under the topic of *network*. This is not a rule-based approach that uses regular expressions or dictionary-based keyword searching techniques. It is an unsupervised ML algorithm to find a bunch of related words and create clusters of text.

You can download this code from https://github.com/dryice-devops/AIOps/blob/main/Ch-8_Anomaly%20Detection.ipynb.

```
#library for mathematical calculations
import pandas as pd
import numpy as np

#library for NLP\text processing
import nltk
from nltk.stem import WordNetLemmatizer, SnowballStemmer
from sklearn.feature_extraction.text import TfidfVectorizer
from sklearn.feature_extraction.text import TfidfVectorizer
from sklearn.decomposition import PCA
from sklearn.preprocessing import normalize
from sklearn.metrics import pairwise_distances
from sklearn.cluster import KMeans
```

```
from sklearn.metrics import silhouette_score,silhouette_samples
from sklearn.manifold import TSNE

import gensim

#library for plotting graphs and charts
import matplotlib.pyplot as plt
import matplotlib
import seaborn as sns
from IPython.display import clear_output

import time
```

Load the file containing some sample events.

```
raw_alerts = pd.read_csv(r'event_dump.csv')
```

Let's understand the data by checking the columns and their sample values.

```
raw_alerts.info()
```

As shown in Figure 8-3, we have 1,050 events (the first row is a header), each with six columns.

```
<class 'pandas.core.frame.DataFrame'>
RangeIndex: 1051 entries, 0 to 1050
Data columns (total 6 columns):
 #   Column            Non-Null Count  Dtype
---  ------            --------------  -----
 0   Source            1051 non-null   object
 1   AlertTime         1051 non-null   object
 2   AlertDescription  1051 non-null   object
 3   AlertClass        1051 non-null   object
 4   AlertType         1051 non-null   object
 5   AlertManager      1051 non-null   object
dtypes: object(6)
memory usage: 49.4+ KB
```

***Figure 8-3.*** *Columns in a dataset*

192

Next let's check the sample events in the data file.

```
raw_alerts.head().transpose()
```

Table 8-1 shows the sample events from the data input file that contains the following slots:

- *Source*: Contains the device hostname/IP

- *AlertTime*: Provides the event occurrence time

- *AlertDescription*: Provides the detailed event message

- *AlertClass*: Provides the class of the event

- *AlertType*: Provides the type of the event

- *AlertManager*: Provides a tool that generates events

***Table 8-1.*** *Sample Events from Input File*

| | 0 | 1 | 2 | 3 | 4 |
|---|---|---|---|---|---|
| Source | AUPRDGLB-HRAPP03 | USPRDEMPRECWEB03 | USDEVGLB-HRAPP02 | UKDEVPAYROLDBA01 | dummy.au-x6.global.com |
| AlertTime | 6/2/2021 03:56:35 | 6/2/2021 05:55:32 | 6/2/2021 07:50:34 | 6/2/2021 07:50:48 | 6/2/2021 10:49:54 |
| AlertDescription | Memory Utilization is 74% in Warning State on... | Memory Utilization is 70% in Warning State on... | CPU Utilization is 95% in Critical State on U... | CPU Utilization is 93% in Critical State on U... | Interface Utilization is 75% in Warning State... |
| AlertClass | OperatingSystem | OperatingSystem | OperatingSystem | OperatingSystem | network performance |
| AlertType | Server | Server | Server | Server | network |
| AlertManager | PlatformMonitoring-Tool | PlatformMonitoring-Tool | PlatformMonitoring-Tool | PlatformMonitoring-Tool | network-monitoring-tool |

Check how many unique `Class` events are present in the data.

```
raw_alerts['AlertClass'].unique()
```

As shown in Figure 8-4, the input dataset contains six different event classes that represent performance and availability issues related to the operating system, network, applications, and configuration changes. These cover most common types of events that get generated in any operations.

```
array(['OperatingSystem', 'network performance', 'Transaction Monitoring',
       'network availability', 'MERAKI', 'configuration'], dtype=object)
```

***Figure 8-4.*** *Unique values in the Class column*

Next, start applying the NLP algorithm. First, we need to download the English words dictionary from nltk. You need to add domain-specific words as that may be missing from the English words dictionary like *app*, *HTTP*, *CPU*, etc. These domain-specific words are important for efficiently creating clusters.

Now apply preprocessing on the event message, which includes removing special characters and punctuation to create tokens. These tokens then pass through the stemming and lemmatization process (explained in Chapter 4) to get more meaningful words that also remove stopwords that represent a set of commonly used words and do not add much meaning to a sentence like *is*, *are*, etc. They carry little useful information and hence can be safely removed.

```python
nltk.download('words')
words = set(nltk.corpus.words.words())
stemmer = SnowballStemmer('english')
def lemmatize_stemming(text):
    return stemmer.stem(WordNetLemmatizer().lemmatize(text,
    pos='v'))

def preprocess(text):
    result = []
    words = set(nltk.corpus.words.words())
    domain_terms = set(["cpu","interface","application",
    "failure",
                        "https","outage", "synthetic",
                        "timeout","utc",
                        "www","simulation","simulated","http",
                        "response","app","network","emprecord",
                        "global_hr","pyroll","employee_
                        lms","demoapp1",
                        "emp_logistics_summary","demo","down",
                        "tcp" ,
```

```
                    "connect","emp_tsms","payroll_sap_gts",
                    "demoapp3","high","state"])
    for token in gensim.utils.simple_preprocess(text):
        if token not in gensim.parsing.preprocessing.STOPWORDS:
            if token.lower() in words or token.lower() in
            domain_terms:
                    result.append(lemmatize_stemming(token))
    return result
```

Next, pass event messages through NLP preprocessing.

```
event_msg = np.array(raw_alerts[['AlertDescription']])
temp = []
for i in event_msg :
    temp.append(i[0])
event_msg = temp

for i,v in enumerate(event_msg):
    event_msg[i] = preprocess(v)

for i,v in enumerate(event_msg):
    event_msg[i] = " ".join(v)
```

For each event sentence, we have the relevant tokens for analysis. Now we need to map these text tokens from the vocabulary to a corresponding vector of real numbers so that we can apply statistical algorithms. This process is called *vectorization*.

You will be using TF-IDF for the vectorization process. TF-IDF is a statistical measure to calculate the relevance of a token. It consists of two concepts: term frequency (TF) and inverse document frequency (IDF).

- *TF*: This considers how frequently the word appears in an event message. As each message is not the same length, it may be possible that a token in a long sentence occurs more frequently compared to a token in a shorter message.

- *IDF*: This is based on the fact that less frequent tokens
  are more informative and important. So if a token
  appears frequently in multiple event messages, like
  *high*, then it's not critical compared to the tokens that
  are not frequent in multiple messages like *down*.

Readers can get more details on TF-IDF at https://en.wikipedia.org/wiki/Tf-idf.

We will be using TF-IDF to vectorize the data in the following:

```
tf_idf_vectorizor = TfidfVectorizer(stop_words = 'english',
                             max_features = 10000)
tf_idf = tf_idf_vectorizor.fit_transform(event_msg)
tf_idf_norm = normalize(tf_idf)
tf_idf_array = tf_idf_norm.toarray()
```

Now we have the final list of tokens on which we can apply the K-means algorithm.

To apply the K-means algorithm, first we need to determine what will be the ideal value of K. For this we will be using the Elbow method, which is a heuristic used in determining the number of clusters in a dataset.

Let's use this method to plot a graph of scores and the number of clusters.

```
number_clusters = range(3, 12)

kmeans = [KMeans(n_clusters=i, max_iter = 600) for i in number_
clusters]
kmeans

score = [kmeans[i].fit(tf_idf_array).score(tf_idf_array) for i
in range(len(kmeans))]
score

plt.plot(number_clusters, score)
```

```
plt.xlabel('Number of Clusters')
plt.ylabel('Score')
plt.title('Elbow Method')
plt.show()
```

From the graph in Figure 8-5 we know that total seven clusters are possible in our dataset. So, let's set the value of K as 7 and execute the K-means algorithm.

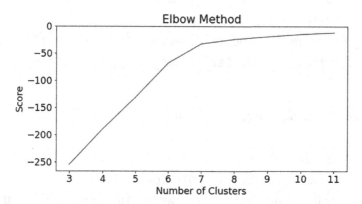

***Figure 8-5.*** *Elbow graph on dataset*

```
df_temp = raw_alerts
no_cluster = 7
kmeans = KMeans(n_clusters=no_cluster, max_iter=600, algorithm
= 'auto')
fitted = kmeans.fit(tf_idf_array)
print("Top terms per cluster:")
order_centroids = kmeans.cluster_centers_.argsort()[:, ::-1]
labels = {}
terms = tf_idf_vectorizor.get_feature_names()
for i in range(no_cluster):
    print("Cluster %d:" % i),
    labels[i]=terms[order_centroids[i, 0]]
```

```
for ind in order_centroids[i, :10]:
    print(' %s' % terms[ind])
```

Now you can see that all the top keywords representing issues are clustered in different clusters.

Let's understand each cluster in more detail by plotting the top issues that are clustered in each cluster.

```
preds = kmeans.predict(tf_idf_array)
df_temp['cluster'] = preds
df_temp['AlertTime'] = pd.to_datetime(df_temp['AlertTime'])
def get_top_features_cluster(tf_idf_array, prediction,
n_feats):
    labels = np.unique(prediction)
    dfs = []
    for label in labels:
        id_temp = np.where(prediction==label)
        x_means = np.mean(tf_idf_array[id_temp], axis = 0)
        sorted_means = np.argsort(x_means)[::-1][:n_feats]
        features = tf_idf_vectorizor.get_feature_names()
        best_features = [(features[i], x_means[i]) for i in
        sorted_means]
        df = pd.DataFrame(best_features, columns = ['features',
        'score'])
        dfs.append(df)
    return dfs

def plotWords(dfs, n_feats):
    plt.figure(figsize=(8, 4))
    for i in range(0, len(dfs)):
        plt.title(("Most Common Words in Cluster {}".
        format(i)), \
```

```
                fontsize=10, fontweight='bold')
        sns.barplot(x = 'score' , y = 'features', orient = 'h' ,
        \ data = dfs[i][:n_feats])
        plt.show()

n_feats = 20
dfs = get_top_features_cluster(tf_idf_array, preds, n_feats)
plotWords(dfs, 10)
```

Figure 8-6 shows that all types of alerts as features get clustered in cluster 0 and their significance or weightage in it. As observed, this cluster primarily consists of CPU-related alerts or issues that occurred with other less significant alerts or issues. As observed from this cluster, CPU utilization alerts are not related to any application-related issues or alerts, which indicates a strong probability that CPU alerts generate a lot of noise, and you may need to tune the CPU utilization parameter.

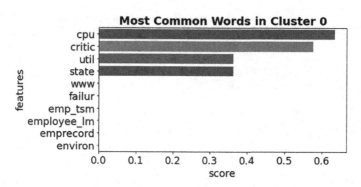

***Figure 8-6.*** *Cluster 0 represents CPU issues*

Cluster 1, as shown in Figure 8-7, represents issues that are primarily related to the application Global HR. Also, this cluster indicates a potential relationship between the Global HR application issue and a memory issue and will help the IT operations teams in the troubleshooting and remediation process. This is quite an important insight that gets automatically detected by ML algorithms without writing any rule.

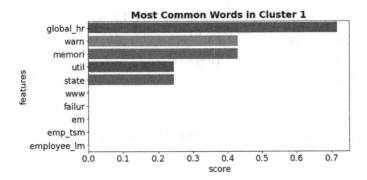

***Figure 8-7.*** *Cluster 1 represents Global HR application issues*

Similar to cluster 1, Figure 8-8 represents issues that are primarily related to the other application, Payroll, which also gets impacted due to memory-related issues. Alerts in this cluster should be used by the application team to understand how to improve its performance and reduce application alerts.

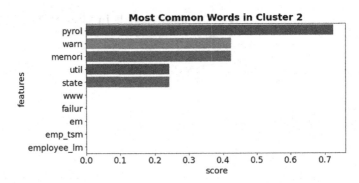

***Figure 8-8.*** *Cluster 2 represents Payroll application issues*

Cluster 3 in Figure 8-9 represents alerts related to the application Employee Record, which needs to be analyzed and fixed. This is also an important input to the problem management team because they need to provide a long-term resolution for this issue.

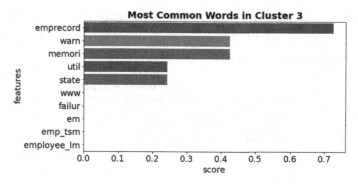

***Figure 8-9.*** *Cluster 3 represents Employee Record application issues*

Cluster 4 in Figure 8-10 represents issues related to network interface utilization that needs to be analyzed by the network team. AIOps can use automation to create a single ticket for the network team against all the alerts that are clustered here. This single ticket will avoid overloading the ticketing system with duplicate incidents.

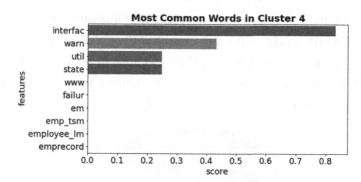

***Figure 8-10.*** *Cluster 4 represents network interface–related issues*

Cluster 5 in Figure 8-11 represents memory-related issues, and we have also observed in previous clusters that memory-related alerts are clustered along with application alerts. This indicates that multiple applications might be using common infrastructure such as a common virtualized infrastructure, which is nearing capacity. Since different clusters are capturing different issues automatically, analyzing them together significantly improves the resolution time.

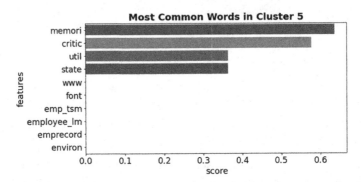

***Figure 8-11.*** *Cluster 5 represents memory utilization–related issues*

Cluster 6 in Figure 8-12 is most interesting as it detected a potential outage. As shown in Figure 8-10, cluster 6 consists of various simulations, transactions, and application-related alerts that indicate some issue at the infrastructure level and not necessarily at the application level. With traditional IT ops methods, finding the root cause in such data would have been difficult, but the ML model created this cluster, which detected a potential relationship and pointed toward the root cause.

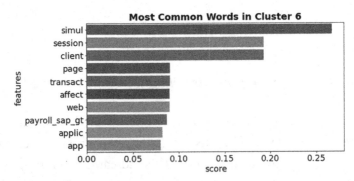

***Figure 8-12.*** *Cluster 6 represents an outage*

Let's observe all the events that are present in this specific cluster, cluster 6.

```
df_appcluster=df_temp[df_temp['cluster'] == 6 ]
df_appcluster.info()
```

As shown in Figure 8-13, there are 43 events that got clustered together in cluster 6.

```
<class 'pandas.core.frame.DataFrame'>
Int64Index: 43 entries, 31 to 1050
Data columns (total 7 columns):
 #   Column            Non-Null Count  Dtype
---  ------            --------------  -----
 0   Source            43 non-null     object
 1   AlertTime         43 non-null     datetime64[ns]
 2   AlertDescription  43 non-null     object
 3   AlertClass        43 non-null     object
 4   AlertType         43 non-null     object
 5   AlertManager      43 non-null     object
 6   cluster           43 non-null     int32
dtypes: datetime64[ns](1), int32(1), object(5)
memory usage: 2.5+ KB
```

**Figure 8-13.** *Cluster 6 events data*

Let's observe the events in this cluster.

```
df_appcluster[['AlertTime','AlertClass','AlertType',
'AlertDescription']]
```

On closely reviewing the events in cluster 6, you can see there are five events from different transaction failures that arrived on July 8 almost at the same time, about 19:30, indicating an outage. As shown in Figure 8-14, one out of five events belongs to a network interface being down, indicating it as a probable cause because immediately after this event, there are events for failed simulated transactions for the application Global HR.

```
df_appcluster[['AlertTime','AlertClass','AlertType','AlertDescription']]
```

| | AlertTime | AlertClass | AlertType | AlertDescription |
|---|---|---|---|---|
| 694 | 2021-07-08 05:45:37 | MERAKI | settingsChanged | Meraki Dashboard settings are changed. [ networkName ] : Raleigh - appliance [ changeDescription ] : {"node_group_management_vlan":{"label":"Management VLAN","old_text":"90","new_text":"100"}} |
| 706 | 2021-07-08 19:28:44 | network availability | network | The current status of Interface GigabitEthernet5/0/3 ◆ DNK-X4 - Gi1/1/1 on the node dummy.us-x1.global.com with IP Address 10.10.10.1 is Down. |
| 707 | 2021-07-08 19:30:44 | Transaction Monitoring | APPLICATION | Simulated transaction Check_Logout_Status failed for Client Session Simulation Global_HR |
| 708 | 2021-07-08 19:31:29 | Transaction Monitoring | APPLICATION | Simulated transaction TSMS_Submit failed for Client Session Simulation Global_HR |
| 709 | 2021-07-08 19:31:54 | Transaction Monitoring | APPLICATION | Simulated transaction Check_Login_Status failed for Client Session Simulation Global_HR |
| 710 | 2021-07-08 19:32:04 | Transaction Monitoring | APPLICATION | Simulated transaction Login_LMS_T02 failed for Client Session Simulation Global_HR |
| 763 | 2021-07-12 20:02:44 | Transaction Monitoring | APPLICATION | Process DEMOAPP2_Process for Application DEMO APP is down. |
| | 2021-07- | Transaction | | Synthetic monitoring failed for Client Session Simulation |

*Figure 8-14.* *Events in cluster 6*

With the help of ML and NLP capabilities, the algorithm discovered and clustered useful information from 1,000+ events, and that was without writing any static rules or using any topology-related details.

Let's also analyze these clusters together over a time scale to uncover more insights.

First let's print the labels that the algorithm has detected based on the events grouped in each cluster.

```
print("Cluster Labels are: \n", labels)
```

```
Cluster Labels are:
{0: 'pyrol', 1: 'cpu', 2: 'interfac', 3: 'global_hr', 4:
'emprecord', 5: 'memori', 6: 'simul'}
```

Now let's observe these clusters over a timeline to explore any useful details.

```
df_temp.plot(x='AlertTime', y='cluster', lw=0, marker='s',
color ="#eb5634",\ figsize=(12,8),rot = 90,  alpha = 0.5,
fontsize = 15,grid=True, legend=False,\ ylabel="Cluster Number")
```

From Figure 8-15, there are few important observations you can make here:

- There is a consistent lot of noise due to CPU utilization events that are represented in cluster 0. This is feedback to tune the thresholds of the CPU monitoring parameter. Or you can submit a recommendation to the capacity planning team to increase the CPU capacity.

- The algorithm automatically created three clusters, one for each application, which primarily contains alerts related to that specific application only. These clusters can be analyzed for incident or problem management-related tasks. These clusters provide a lot of visibility to the applications teams and are immensely helpful in executing service improvement programs.

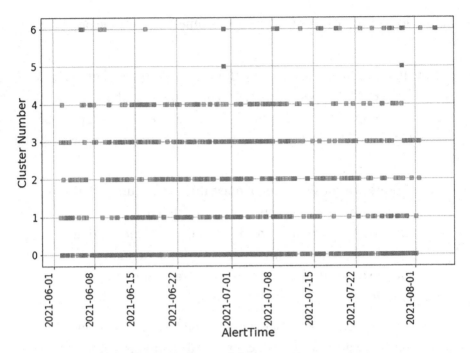

***Figure 8-15.*** *Timeline analysis of cluster*

This AIOps use case of anomaly detection also makes it extremely helpful and efficient to detect potentially critical events, such as security events, from the enormous pile of events and initiate automated actions, such as invoking a kill switch to minimize the impact. You have seen how K-means clustering detected noise and can generate recommendations for problem and capacity management processes.

---

**Note**    As per best practices, security alerts should not be integrated with the operations event console. Security alerts are exclusively for the eyes of security experts and not for everyone. Ideally, there should be a separate SIEM console for the security operations team to monitor.

---

Though a K-means algorithm is simple to implement even with a very large dataset, it has a few challenges from an AIOps perspective.

- The number of clusters plays a crucial role in the K-means algorithm's efficiency.

- The K-means cluster gets impacted due to the presence of noise or a high number of outliers in the dataset. Centroids get stretched to incorporate outliers or noise.

- A higher number of dimensions (parameters or variables) has a negative impact on the efficiency of the algorithm.

Along with anomaly detection, another common use case of AIOps is correlation and association, which will be discussed next.

# Correlation and Association

Correlation is a statistical term that means identifying any relationship or association between two or more entities. From an AIOps perspective, correlation is used to determine any dependency or relation between entities and cluster them together for efficient analysis. We covered multiple regression algorithms that determine the relationship between multiple entities to establish correlation in Chapter 5.

However, establishing correlation at the event layer is a bit difficult as multiple events from varied sources come at a varied time interval. A time-based sequence or pattern of event is quite rare at the event layer. One of the algorithms that works efficiently at the event layer to determine correlation and association is DBSCAN, and it is particularly useful in scenarios where data points are close to each other (dense) along with the presence of a considerable amount of noise or outliers.

# Topology-Based Correlation

The accuracy of clustering is one of the most important KPIs of an event management system that drives the efficiency of the engine. To improve the accuracy of clustering and the determination of the root cause, you need to provide some topology context. But the CMDB and discovery are challenges in themselves, and having a 100 percent accurate CMDB with 100 percent discovery and modeling of infrastructure is a tough task.

Considering the various challenges with the CMDB and discovery, topology correlation can still be implemented for network topology correlation as well as for application topology correlation. Consider the sample topology shown in Figure 8-16. Connections marked in black represent the network topology, whereas connections marked in red represent the application topology. We will use this sample topology to discuss correlation for network topology and application topology.

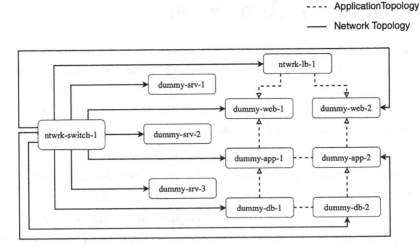

***Figure 8-16.*** *Sample topology diagram*

# Network Topology Correlation

This correlation uses the network connectivity diagram as the topology to perform correlation. For example, if a core switch connecting hundreds of servers and devices to its interfaces is down, then alerts from this switch are causal, and all other alerts from the underlying network devices and servers as well as applications running over them are considered impacted.

The IT operations team can focus on the causal CI that is the network switch instead of manually filtering alerts in the system due to this outage. This correlation needs to determine the starting position and then depth of hierarchy to compare and create this topology and arrive at the causal event. Since storing the entire topology or discovering everything may not be practical feasible, the next best approach is to store a subset of topology information for critical devices or application at the event management layer for correlation and every 24 hours (or weekly if the environment is not very dynamic) updating it using automated discovery/scripts or manually. The network topology for the sample topology in Figure 8-4 can be stored as a key-value pair at the event manager layer, as shown here:

```
{"network_topology": [
     {"child-ci": "dummy-app-1 ", "parent-ci": " ntwrk-
     switch-1 "},
     {"child-ci": "dummy-app-2 ", "parent-ci": " ntwrk-
     switch-1 "},
     {"child-ci": "dummy-web-1 ", "parent-ci": " ntwrk-
     switch-1 "},
     {"child-ci": "dummy-web-2 ", "parent-ci": " ntwrk-
     switch-1 "},
     {"child-ci": "dummy-db-1 ", "parent-ci": " ntwrk-
     switch-1 "},
```

```
{"child-ci": "dummy-db-2 ", "parent-ci": " ntwrk-
switch-1 "},
{"child-ci": "dummy-srv-1 ", "parent-ci": " ntwrk-
switch-1 "},
{"child-ci": "dummy-srv-2 ", "parent-ci": " ntwrk-
switch-1 "},
{"child-ci": "dummy-srv-3 ", "parent-ci": " ntwrk-
switch-1 "},
{"child-ci": "ntwrk-lb-1", "parent-ci": " ntwrk-
switch-1 "}
]}
```

Whenever alerts are generated, the event management algorithm
will use this data to determine relationships, perform correlation, and
determine the causal and impacted CIs. This method is dependent on the
static information, and thus it has its limitations and challenges. Storing a
high volume of topology data can severely impact the event management
tool performance.

From an AIOps perspective, it's recommended to offload this
correlation at the network monitoring layer, which has a native capability
to scan the subnet ranges, discover devices, and configure them in
availability monitoring using ICMP or SNMP. Network monitoring tools
have real-time visibility on the availability of all CIs, and their topology, as
well as can dynamically update them based on changes in the network.
A network monitoring tool should perform this correlation using the
topology detected, filtering out impacted CI alerts and forwarding
causal alerts to the event management layer. If needed, the network
monitoring tool can be configured to forward the impacted CI alerts with
the information severity, whereas a causal alert can be forwarded with a
critical severity.

In cases of an open source network monitoring tool with limited to no capability for native topology-based correlation, organizations can still import the required topology data in the event management layer and perform correlation as discussed earlier.

# Application Topology Correlation

This correlation leverages the application architecture with a relationship between application components. This can be as simple as a vCenter farm running on multiple hypervisors in a cluster and hosting various VMs. This can also be a more complex application architecture for a business application like an insurance claim processing system running on multiple nodes hosting web servers, DB servers, EDI systems, and SAP systems.

Unlike network topology correlation where only the availability of systems are considered, application correlation considers the availability as well as the performance of applications. This correlation not only needs details of network topology but also the details of application blueprints and key performance indicators. Consider an example where an alert is received about database processes being down, which further causes alerts for the following:

- High request in application queue from application servers

- High response time of an application URL from web servers

- Timeout alert from synthetic monitoring from the load balancer

At this point of time, many other unrelated alerts may also be received at event management. In this case, all systems are up and running, but the issue is at the application level. Here the application-level correlation helps in isolating faults.

Application architecture or blueprints ideally should be available (or discovered) and stored in the CMDB as service models. But organizations face lots of challenges in maintaining these blueprints because of the following three most common reasons:

- Discovering application topology requires domain knowledge to determine the pattern that needs to be detected in discovery. Discovery tool SMEs don't possess such vast domain knowledge.

- Information is federated and available inside different tools managed by multiple teams.

- There are challenges in getting clearance to read application logs/services/processes. Logs usually contain PII/PHI/confidential data, which causes a red flag to be raised by the security and application teams.

From an AIOps perspective, whenever alerts related to an application arrive, then algorithms should leverage application topology to perform the necessary correlation. Considering these challenges with the CMDB, AIOps algorithms can be used to learn patterns and dependencies and automatically derive topology. In our sample topology example, algorithms can use the following features to understand topology and perform application correlation:

- Learn by analyzing the similarity of hostnames/IP addresses as servers belong to one application usually share a common prefix in the hostname or common subnet details.

- Learn by analyzing the arrival time of the events and detecting patterns from it.

- Learn by analyzing message texts in the event, whether it belongs to the availability category or the performance category or if there any common application/service name in events that comes together within a short interval.

- Learn by analyzing the event class and source. Data from multiple operational management databases (OMDBs) like vCenter, SAP, SCCM, etc., is quite useful in this learning as these are the sources where different CIs are getting managed. For example, the vCenter database provides the data of the mapping of VMs to the ESX farm where the event source will be vCenter and the class will be Virtualization. Similarly, SAP can provide application mapping details where the source will be SAP System and the class will be Application. The algorithm can use this data to correlate events and find out whether the issue is due to the underlying hypervisor or due to application performance.

- Like network topology correlation, you can store the application topology data for critical applications at the event management layer and let the system learn from it. Unless the application is running on the cloud or on SDI, the application topology usually does not get updated as frequently as the network topology data, so it can be synced on a monthly or quarterly basis. For cloud and SDI-based applications, it is comparatively easy to use native APIs and update the topology data as soon as a change happens.

```
{"application_topology": [
     {"child-ci": " dummy-db-1 ", "parent-ci": " " ,
     "application":"dummy_app","kpi":["type":"process","na
     me":" dummy-process"]},
     {"child-ci": " dummy-db-2 ", "parent-ci": " " ,
     "application":"dummy_app","kpi":["type":"process","na
     me":" dummy-process"]}
     {"child-ci": "dummy-app-1 ", "parent-ci": " dummy-db-1",
     "application": "dummy-app","kpi":["type":"service","na
     me":" app_service "]},
     {"child-ci": "dummy-app-2 ", "parent-ci": " dummy-db-2" ,
     "application":"dummy_app","kpi":["type":"service","na
     me":" app_service "]},
     {"child-ci": "dummy-web-1 ", "parent-ci": " ntwrk-lb-1" ,
     "application":"dummy_app","kpi":["type":"url","name":"
     dummy-url.com "]},
     {"child-ci": "dummy-web-2 ", "parent-ci": " ntwrk-lb-1" ,
     "application":"dummy_app","kpi":["type":"url","name":"
     dummy-url.com "]},
     {"child-ci": " dummy-web-1 ", "parent-ci": " dummy-
     app-1 "     ,"application":"dummy_app","kpi":["type":"url
     ","name":" dummy-    url.com "]},
     {"child-ci": " dummy-web-2 ", "parent-ci": " dummy-app-2 " ,
     "application":"dummy_app","kpi":["type":"url","name":"
     dummy-    url.com "]}
]}
```

Assume if the database process running on dummy-db-1server goes down, it will generate a database alert as well as an application queue alert from dummy-app-1 and URL response time alert from dummy-web-1 servers. Using the previous data, the event management system can create clusters and correlate these alerts as well as highlight which one is causal and which ones are impacted.

The availability of application topology or service models can definitely increase the accuracy of algorithms, but it is not a must-have requirement from AIOps. Both supervised and unsupervised machine learning algorithms showed lot of potential and can help a lot in performing the application topology correlation.

# Summary

This chapter covered the important use cases in AIOps around anomaly detection using K-means clustering. You also used NLP techniques such as stopword removal and TF-IDF to make sense of textual data. You learned about other techniques such as application dependency mapping and the CMDB and how they are relevant in AIOps. In the next chapter, we will cover how to set up AIOps as a practice in an organization.

# CHAPTER 9

# Setting Up AIOps

This chapter provides best practices for the AIOps journey and gives guidance on setting up AIOps in your organization.

Having learned about AIOps, its techniques, and its benefits, now it is time to look at how to implement AIOps in an organization. Figure 9-1 describes the steps for setting up an AIOps practice in an organization.

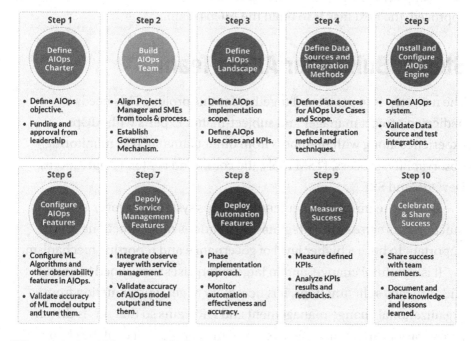

*Figure 9-1.* *AIOps establishment process*

Let's begin the AIOps journey by defining the AIOps charter.

© Navin Sabharwal and Gaurav Bhardwaj 2022
N. Sabharwal and G. Bhardwaj, *Hands-on AIOps*,
https://doi.org/10.1007/978-1-4842-8267-0_9

# Step 1: Write an AIOps Charter

AIOps is not just a technology change; it is a cultural change that involves changes to processes and people skills, so it needs commitment from executive management for it to be implemented successfully.

AIOps impacts various functions and teams including process teams, process owners, the command center, and IT operations. There may also be other initiatives in the organization such as site reliability engineering and DevOps that need alignment with the AIOps initiative.

Having a formal project charter, funding, and approval of executive leadership so that the AIOps project can get the required funding and attention is essential. This will help get the buy-in from various teams that are going to be impacted by the project. Once the AIOps charter gets approved, the next task is to build the AIOps team.

# Step 2: Build Your AIOps Team

The next step is to get a team together for this project. This needs a dedicated project manager and subject-matter experts with AIOps experience along with members from other teams such as monitoring, observability, process, ITSM tools, the command center, IT operations, DevOps, and SRE.

The core team will implement the AIOps system; however, since this function cuts across and integrates with various other functions, it is important to have a higher level of governance and reporting mechanism.

It is a cultural change, and in large enterprises with siloed hierarchies and vertically split functions, it is necessary to get representation from organizational change management and HR teams so that this is driven across various functions effectively and the people aspect of the change is handled through a process-oriented approach. At this stage, you need to start exploring and evaluating the scope and goals for an AIOps implementation, which we will be covering next.

# Step 3: Define Your AIOps Landscape

Before embarking on any technology change, it is essential to define what the goals are and what you trying to achieve. The team should go through various areas in AIOps and identify what would work for them best.

The next step should be to either define a subset where AIOps will be implemented or define a phased approach to implementation where various functions and features will be rolled out in stages.

Thus, you can go ahead and deploy the complete suite of AIOps features for a specific business line or go step-by-step and implement module by module for the entire enterprise as a whole. The decision should be based on the organizational structure, team structure and size, scale, and complexity. For a highly siloed and large organization, it would be better to implement it in a business vertical, whereas a medium-sized organization can proceed to implement it for the entire organization.

The first step in the planning stage is to gather the data around the current implementation of various tools and technologies and the ITSM processes. This should include the following:

- Current monitoring tools landscape

- Current infrastructure and application landscape

- Data sources for topology/CMDB

- Service management tools

- Processes

- Command center function and procedures

- Resolution groups and procedures

- Current issues and challenges in monitoring and management

- Rule-based policies in existence for event correlation

- Automation coverage and tools

After gathering all of this data, it is necessary to have a tollgate at this stage. The analysis of this data will result in better understanding of the maturity of the organization in terms of monitoring and service management. It is possible that the analysis of this data may lead to another subproject where some of the basic elements of monitoring or service management need to be enhanced or changed so that the AIOps project gets the integrations and data for it to function.

If the analysis results in another subproject to enhance the monitoring and management tools or processes, it should be handled as a separate project under the same umbrella program since the owners for this project may be different. The AIOps project can continue on its journey while this monitoring enhancement project runs in parallel.

As part of the data gathering, you should gather data on the current KPIs so that before and after implementation of AIOps the KPIs can be compared. The following KPIs are important indicators for measuring the success of AIOps deployment and should be measured before implementation and on an ongoing basis after deploying AIOps.

- Alert to incident ratio

- Mean time to respond

- Mean time to resolve

- SLA metrics for P1 and P2

- Time for closure of SRs

- Percentage of SRs automatically resolved

- Percentage of incidents automatically resolved

- Availability of critical systems

The next step in the AIOps journey is to define data sources and their integration.

# Step 4: Define Integrations and Data Sources

After having collected the relevant data for the environment, you will have clarity on what integrations are required. The integrations will be broadly with the following systems:

- Monitoring tools

  - SNMP (NetBackup, SAP Solman, etc.)

  - Syslog (config change alerts, UNIX kernel alerts, etc.)

  - APIs (Zabbix, vCenter, etc.)

  - Other database-based connectors

- Service management tools using APIs

- CMDB using APIs

- Knowledge management sources using APIs

- Automation tools using APIs

In general, you will find multiple monitoring tools in the environment. Some of them will support APIs, and the data can be ingested into AIOps using API-based connectors. Tools like vCenter provide APIs to check the status of the underlying visualized infrastructure; others may send SNMP-based alerts. For example, SAP Solman triggers SNMP alerts for SAP resources. There are multiple options for getting the data into the AIOps engine; these should be evaluated and then implemented based on the best approach.

Typically, you will see network monitoring, server monitoring, and application monitoring tools in the environment along with specialized tools for monitoring backups, jobs, storage, and other OEM devices. All these need to be integrated with the AIOps engine so that all the monitoring data resides in the single engine for analytics and for its algorithms to be trained.

From a service management perspective, there will be typically one tool, and the CMDB may be within the same toolset along with the KEDB and basic knowledge management. Most of the leading tools provide APIs, so data can be easily ingested using API-based connectors for these systems.

Automation tools may get integrated based on use cases that you are planning for AIOps and whether you plan a direct integration with AIOps or through service management tools.

Even before the AIOps engine gets trained, once integrations are complete, the organization will start seeing the benefits of a single-pane-of-glass view into operations with all alerts in a single console.

This stage concludes the important step in the AIOps journey where prerequisites get completed and you actually start the installation, deployment, and configuration steps, which will be covered in subsequent sections.

# Step 5: Install and Configure the AIOps Engine

Once all the data sources and data has been identified, the next step is to start installing the AIOps engine. This depends on the tools and technology that you are choosing. Since these are complex systems, organizations generally take implementation services or professional services from partners who are well versed with the product and have the required expertise to implement these solutions. The Core AIOps team can shadow the implementation and learn and gain expertise while the AIOps engine is getting implemented. Alternatively, if the required skills and expertise are available in house, the team can start building the solution themselves.

For large, complex deployments, it's highly recommended to proceed with a phased approach to ensure minimal to zero disruption to the services or business. Trying to accomplish everything quickly in minimal time needs good experience and multiple stakeholders' support. In the first or initial phase of implementation, install the base AIOps solution and select and integrate APM (transaction monitoring), platform, and network data sources based on the output from step 3 and step 4 as explained earlier. These integrations will provide visibility into the health of servers, databases, and network devices, which covers the majority of organization estate.

This phase provides good learning and ample confidence for subsequent phases. Learning from phase 1 can then be used for improving the existing integrations as well as executing other data sources integrations like APM (deep dive), storage, backup, VMware, SAP Solman, job scheduling, etc., in subsequent phases. It is important to note that selecting data sources for integration in a specific phase depends on the criticality and urgency of the organization's requirements.

Configuring the AIOps engine with all data sources integrations as defined in step 3 and step 4 may take time depending upon their coverage and maturity and organizational processes. For example, vCenter is planned to be integrated, but not all ESXs and farms are not configured in it, or an upgrade of storage, like EMC, comes during its integration with the AIOps engine. It is practically difficult to have detailed insight into all technology towers, especially in large organizations, and that's why continuous feedback is important to tune the integrations and expand the coverage.

To reduce complications and have better time to value (TTV), SaaS platforms can be leveraged. On SaaS platforms, it would not involve installation, so you can proceed directly with configuring the solutions once the AIOps vendor onboards you onto the platform.

Once the AIOps engine is set up and configured, the next step is to set up and test the integrations. This involves getting the integration adapters configured and integrated. Each data source needs to be tested once the integration is set up to validate that the required data is being captured correctly in the AIOps engine.

If you are directly integrating with automation, then the outbound integration to automation is also set up and configured.

At this stage, your AIOps system is set up and ready for validation by the operations teams where they can see all the events and alerts in a single console. You can proceed to configure roles, users, and access based on roles and provide them with the required console and dashboards.

At this stage, your AIOps engine acts like a central repository of all the data sources but doesn't do any AIOps functionality and features for event management.

# Step 6: Configure AIOps Features

The next step is to use the core AIOps features of the platform to configure the event management functionality. We have already covered the various layers in event management in earlier chapters; these need to be implemented and configured in the system. For large and complex deployments, this step happens in parallel with the integration phases of step 5. Functions such as deduplication, enrichment, topology correlation, application correlation, anomaly detection, etc., process the events coming from integrations with various data sources and select the qualified alerts for incident creation and subsequent automation of diagnosis and remediation. Again, continuous feedback is important while configuring the AIOps features and functions to tune the performance and accuracy of the AIOps engine. We have discussed in detail the implementation and best practices around these AIOps functions in previous chapters.

After configuring all the features, the system starts to learn from the data and the feedback being provided by the operations teams and forms a continuous loop where new data and new feedback provides new inputs to the AIOps engine to fine-tune itself and improve its coverage and accuracy.

Once you have set up everything and the system is up and running, you should monitor the system on an ongoing basis for accuracy and any drift that may happen in the data. During its lifetime there will be new integrations and new tools that may come up in the environment that need integration with the AIOps engine. These need to be handled as and when new monitoring capabilities are brought into the environment.

# Step 7: Deploy the Service Management Features

You can start using the capabilities of AIOps in the service management space next. As defined earlier in the Engage layer, the step-by-step approach can be implemented, and each function or feature can be rolled out as part of the plan. The detailed activities in AIOps service management will start with the integration of the Observe layer with incident creation and assignment, and the remaining features in the Engage layer will get implemented in order. An organization that prefers a highly cautious approach can proceed with selective automatic incident creation, such as creating automatic incidents only for production servers or only for a down event, etc., and then gradually expand automatic incident and auto-assignment for other types of qualified alerts, continuously improving both mean time to respond and mean time to repair.

# Step 8: Deploy Automation Features

Once the AIOps engine has been configured for the Observe and Engage phases the automation actions can be configured.

The automation engine needs to be configured in two phases. Phase-1 can be referred as "human assisted automation", where the AIOps engine provides probable cause that needs to be validated by the human agents and the automation engine provides the recommendation for resolution that also needs validation by the human agent.

Once the AIOps Observe engine has run its course over a few weeks and reached a level of confidence score and once the probable cause can be considered as the root cause and the confidence score of the automation engine has also reached a threshold based on the feedback from human actions, we can proceed to deploy the system in fully automated mode.

From here on, the system will gradually adding more root cause and automations to its repository to cover more areas under full automation mode.

Ensure that all the modules are monitored for accuracy and any drift that may happen is tracked and remediated for the accuracy levels to be maintained. New automation runbooks and tools may get deployed in the environment and will need integrations as and when they are available in the environment.

Once the AIOps system is set up and ready, it is time to observe and measure its value and benefits.

# Step 9: Measure Success

Once the implementation has been rolled out, it is time to measure where you stand. In the initial stages of the project, you measured the initial KPIs when you were embarking on the AIOps journey. It is now time to measure these KPIs again and see where you stand after implementation. Remember, we had defined the following KPIs:

- *Alert to incident ratio*: You should see a significant improvement in the alert incident ratio. Because of the removal of noise in the system and probable cause analysis, false alerts that trigger an incident do not happen or are significantly reduced. This also results in a lower number of incidents in the incident management process since false or duplicate incidents are now suppressed by the AIOps engine.

- *Mean time to respond*: The mean time to respond is significantly reduced since the automated analysis of events provides better input to the response teams and some of the response is automated by the system as well. This metric should see a marked improvement.

- *Mean time to resolve*: This metric should also see significant reduction since the time to respond is reduced, and the time to resolve problems through automation is significantly lower. Less time is spent in trying different options, and the AIOps engine is able to guide the resolution teams to the right runbook for resolving the issues. This metric also impacts the availability of the systems, and you should see an improvement in the availability of the applications and infrastructure.

- *SLA metrics for P1 and P2*: Since the SLAs are based on response and resolution, a marked improvement on these would translate to better scores on SLAs. In fact, the operations teams can go back and commit on better SLAs because of improvement in operations using AIOps.

- *Time for closure of SRs*: Service requests get automated as a part of AIOps initiative, and thus the time required for closure of SRs is significantly reduced. There are also fewer errors since the SRs are handled automatically, thus avoiding human errors.

- *Percentage of SRs automatically resolved*: Since automation resolves many SRs, this number sees significant increase through automation.

- *Percentage of incidents automatically resolved*: Once automation has been established as part of AIOps toolset, then automated incidents significantly increase.

- *Availability of critical systems*: A high degree of automation results in lower time to diagnose and resolve incidents, and this results in the higher availability of critical systems.

# Step 10: Celebrate and Share Success

Once you are done with all the layers, the final step in your AIOps journey is sharing your success with the business, leadership, and all the teams. Prepare a detailed case study leveraging the metrics that you set out to achieve and the actual results. Cover how you undertook the journey and what challenges were faced and how those were overcome. Share the knowledge with the AIOps community so that others can learn from your experiences. You can drop a note to us on your journey on AIOps as well at Feedback@AgileInfraOps.com.

Next, let's discuss some best practices and guidelines for AIOps implementation.

# Guidelines on Implementing AIOps

The following are some guidelines when implementing AIOps.

## Hype vs. Clarity

Do not undertake an AIOps project or any other project for that matter because it is fancy and hyped. There should be a definitive need and a use case for deploying AIOps. Clarity of need and purpose is essential for your successful journey to AIOps journey.

## Be Goal and KPI Driven

It is important that you measure your KPIs in the beginning and the end of the implementation. There are implementations that will fail because there is no clear idea on what the team wants to achieve. Having KPIs and knowing how they get impacted by the current project keeps everyone focused on the end outcome.

## Expectations

AIOps applies extensive automation and statistical analysis to the events, performance metrics, logs, and trace data collected from monitoring tools to learn behaviors, identify anomalies, correlate alerts, reduce noise, and pinpoint root causes.

One has to understand and accept that machine learning technologies are probabilistic in nature, and thus they cannot be 100 percent accurate all the time. The idea is to progressively train them to a level of accuracy and confidence where the recommendations can be used by the IT operations teams in finding and resolving problems.

# Time to Realize Benefits

Have realistic expectations for the time that machine learning needs to analyze data, build and train models, and begin providing insights, such as performance anomalies, grouped alerts, and root causes. For example, identifying weekly seasonality requires at least a couple of weeks of observation.

Given the time and feedback, AIOps will provide IT operations with more accurate insights, allowing better decisions to be made. However, expecting AIOps to be a turnkey solution that automates everything on day one is an unrealistic expectation.

# One Size Doesn't Fit All

Every organization is unique, and you will have differences in infrastructure, application landscape, monitoring, and management tools, and hence there will be differences in approach, implementation, and the time to realize value. The size, scale, and complexity of an environment also have a bearing on the time taken to implement and realize value. However, we have seen that AIOps, if implemented correctly, provides rapid value realization and positive ROI.

# Organizational Change Management

Familiarity with AIOps tools and processes is one part of the puzzle; however, a bigger piece is how to get buy-in from different teams that need to be part of this initiative; thus, organizational change management is a key factor in successful AIOps projects.

# Plan Big, Start Small, and Iterate Fast

Rather than attempting to do everything in one massive undertaking, start small. That gives you the chance to learn from your accomplishments, validate and fine-tune your approach, acquire and build on capabilities, and achieve all-important quick wins for your organization. Taking on too much at once can lead to disappointments and may lead to poor business adoption.

# Continually Improve

Once it's deployed, you can continuously improve by extending the coverage and scope of AIOps and fine-tuning its algorithms to gain higher accuracy. AI is a dynamic and fast-evolving field, and newer technologies in AI are emerging that can impact how AIOps will be delivered in the near future.

# The Future of AIOps

With advancements in AI and ML technology, AIOps systems will also get enhanced with more accurate predictions. Let's understand the potential future of AIOps in enterprises.

Observability has a key role to play in DevOps and site reliability engineering. With new tools and techniques available for observability, the data available for analysis increases multifold. With AIOps and observability, the operations teams will have much higher visibility and greater control over their infrastructure and application landscape. After implementing AIOps in the observability area, deploying AI techniques to make sense of the event data, and helping the operations teams find the root cause of issues, the next step is to leverage these techniques to automate the resolution itself.

Tools such as iAutomate provide these capabilities where advanced machine learning techniques are used to take the probable cause as an input and apply AI to find the right automation and take automation in the operations area to the next level of maturity. Currently enterprises that are early adopters are using these technologies in the monitoring and observability domain and implementing automation to realize the complete benefits of end-to-end automation.

The next frontier for AIOps is the better integration of the development pipeline and using these technologies in the development area. AIOps should be able to help by automating the path from development to production, predicting the effect of deployment on production and responding to changes in the production environment.

With cloud computing and the adoption of microservices architecture, the resolution of root cause, especially the root causes that are related to capacity or performance of the infrastructure, would simply be to spin more containers or to launch more virtual machines. Thus, microservices architecture on one hand increases the complexity of the application by breaking it down into many services to map and manage; however, it eases the task of resolution for performance and capacity issues by providing an easy automation capability to spin up more containers quickly and handle the spike in workload.

AIOps and DevOps together solve a lot of problems. DevOps brings together the teams that were siloed earlier, while AIOps brings together data from multiple sources at a single place for insights and analytics. Both require a level of cultural change since they look at the entirely of the process as well as cut across multiple teams and processes. We will see much better integration of AIOps and DevOps in times to come with enterprises making the transition to AIOps by relooking at their DevOps processes and vice versa. Organizations will go on to create interfaces between the AIOps and DevOps processes and create procedures that cut across the two domains.

AIOps in its current form and shape will continue to be adopted by enterprises, and with better integration, DevOps organizations will start using the AI techniques in other areas beyond operations. AIOps is primarily focused on the operations world today; however, the tools and techniques and the algorithms are equally applicable in the development world. On one hand, AIOps data will be used as an input into the development processes for creating more resilient and optimized applications, and on the other hand new use cases will evolve for taking AIOps technology into the development world.

Similar to the operations world, where we have multiplicity of monitoring and management tools, the development world also has multiplicity of tools. The development tools that generate the development data are the Agile dashboards that track the pipeline of features to be built, the team statistics, the burndown charts, and the schedule of releases. Then there is data from testing tools that include functional testing, performance testing, code quality, and security and vulnerability testing. All this data is still analyzed by humans, and the decisions for deployment are made by the release and management teams. There are enterprises that have perfected the art of continuous delivery and continuous deployment; however, using AIOps technologies to go through the development data and provide insights and analytics similar to what is now available in the operations world will be the next frontier for AIOps technologies.

Some application performance management tools have started using these technologies to provide insights into the application code to find where the root cause of performance or availability issues are. Though most APM technologies are still rule or topology based, a few vendors have moved in this direction to bring machine learning and AI technologies to the area of application development and testing.

There will be better integration of the AIOps technologies in the development pipeline, wherein events and data getting generated through the pipeline are sent to the AIOps pipeline for analysis and automated actions.

The next step is where AIOps is able to provide intelligent guidance on the code and configuration changes being made in various environments. This can be based on the data being generated by the pipeline. There would be integration with historical analysis of issues and bugs that were generated, and these can be source of input to arrive at predictions on the current pipeline.

There are lot of false positives that are generated in testing. The automated testing tools that look for code quality or static code analysis often generate false positives. AIOps can be used to eliminate these false positives. Regression testing can be automated and simplified through the use of machine learning technologies. Thus, AIOps technologies in the DevOps pipeline can reduce the amount of testing, help in automated testing, reduce false positives, and guide the team in focusing their time and energy on the most relevant pieces.

AIOps can be intelligently used in both preproduction and production systems to analyze the behavior of applications and impact of configuration changes on the application performance. These alerts can be configured to map the changes between the preproduction and production deployment to find any issues that may crop up because of configuration issues or because of changes in the workload pattern.

AIOps is already being used by advanced automation systems to find the best remediation available and to apply it automatically; however, future systems will also enable ChatOps-based collaboration between the developers and operations teams to enable knowledge sharing and just-in-time knowledge retrieval for bug fixing as well as fixing errors in production environments. Thus, AIOps will see its usage expand beyond operations into the entire DevSecOps value chain.

Though there has been work in the areas of predictive analytics, there will be further enhancements to the algorithms and techniques being used in AIOps to deliver ever-increasing use cases and also to improve the accuracy of the results. As more and more data is fed into the systems, the AIOps engines will become more powerful and provide ever deeper insights and analytics.

The core algorithms that involve lots of machine learning models today will evolve to use deep learning technologies that will provide better insight since the amount of data available to AIOps systems will increase, and neural network systems will become more accurate.

AIOps has had a great start, and organizations are seeing immense benefits of using these technologies today. It will go on to conquer new domains and areas beyond IT operations and eventually impact the entire IT value chain.

# Summary

In this chapter, we covered a step-by-step approach to implementing AIOps in an organization. We also covered what to watch out for while implementing AIOps and avoid pitfalls. We gave guidance on a successful implementation of a project and ways to measure success as well as the important KPIs to measure the success of an AIOps project.

This brings us to the end of this book; we have covered AIOps in depth, beginning with the definition and areas that AIOps covers to a complete architecture of AIOps covering the Observe, Engage, and Act phases. We also covered various machine learning tools and techniques and algorithms that can be used in the AIOps domain. We gave you a detailed walk-through of the process that you can use to implement AIOps in your organization in a step-by-step manner. We sincerely hope that you enjoyed this book as much as we enjoyed authoring it. We look forward to your feedback and comments at feedback@AgileInfraOps.com.

# Index

## A

Act phase
  analytics, 56
  change orchestration, 55–57
  collaboration, 57
  execution, 51, 52
  feedback, 57
  incident resolution, 52
  ITSM system, 50
  recommendation, 50
  scheduling/conflicts, 55
  service request
    fulfilment, 53, 54
  technical execution, 49
  visualization, 57
Akaike information criteria
  (AIC), 172
Anaconda installation process,
  114, 115
Anaphora resolution, 105
Anomaly detection
  correlation/association, 207
  application
    components, 211–215
    network connectivity
     diagram, 209–211
    topology, 208, 209
  description, 187

K-means (*see* K-mean
  algorithms)
  unusual scenarios, 187
Application discovery/insights, 58
Application topology correlation,
  208, 211–215
Architectural view
  act phase, 49–58
  big data (*see* Big Data
    technologies)
  core services, 19, 20
  engage area
    agent analytics, 44
    analytics, 44
    change analytics, 45
    collaboration, 47, 48
    elements and stages, 42
    feedback, 48
    incident creation, 42
    knowledge management, 48
    process analytics, 45, 46
    service level
     management, 45
    service management, 41
    task assignment, 43, 44
    visualization layer, 46
  historical data, 20
  key areas, 23

Printed in the United States
by Baker & Taylor Publisher Services